Technology Manual

Dorothy Wakefield
University of Connecticut

Kathleen McLaughlin
Manchester Community Technical College

Beverly Dretzke
University of Minnesota

FOURTH EDITION

Elementary Statistics
Picturing *the* World

Larson | Farber

PEARSON

Prentice Hall

Upper Saddle River, NJ 07458

Editorial Director, Mathematics: Christine Hoag
Editor in Chief, Mathematics and Statistics: Deirdre Lynch
Editorial Assistant: Joanne Wendelken
Senior Managing Editor: Linda Behrens
Project Manager, Production: Robert Merenoff
Art Director: Heather Scott
Supplement Cover Manager: Paul Gourhan
Supplement Cover Designer: Victoria Colotta
Operations Specialist: Ilene Kahn
Senior Operations Supervisor: Diane Peirano

© 2009 Pearson Education, Inc.
Pearson Prentice Hall
Pearson Education, Inc.
Upper Saddle River, NJ 07458

Pearson Prentice Hall™ is a trademark of Pearson Education, Inc.

The author and publisher of this book have used their best efforts in preparing this book. These efforts include the development, research, and testing of the theories and programs to determine their effectiveness. The author and publisher make no warranty of any kind, expressed or implied, with regard to these programs or the documentation contained in this book. The author and publisher shall not be liable in any event for incidental or consequential damages in connection with, or arising out of, the furnishing, performance, or use of these programs.

Printed in the United States of America

10   9   8

ISBN-13:   978-0-13-601308-2

ISBN-10:   0-13-601308-2

Pearson Education Ltd., London
Pearson Education Singapore, Pte. Ltd.
Pearson Education, Canada, Ltd.
Pearson Education—Japan
Pearson Education Australia Pty, Ltd.
Pearson Education North Asia, Ltd.
Pearson Educación de Mexico, S.A. de C.V.
Pearson Education Malaysia, Pte. Ltd.
Pearson Education Upper Saddle River, New Jersey

The MINITAB Manual

Dorothy Wakefield
University of Connecticut

Kathleen McLaughlin
Manchester Community Technical College

FOURTH EDITION

Elementary Statistics
Picturing *the* World

Larson | Farber

PEARSON

Prentice Hall

Upper Saddle River, NJ 07458

▶ Introduction

The MINITAB Manual is one of a series of companion technology manuals that provide hands-on technology assistance to users of Larson/Farber *Elementary Statistics: Picturing the World.*

Detailed instructions for working selected examples, exercises, and Technology Labs from *Elementary Statistics: Picturing the World* are provided in this manual. To make the correlation with the text as seamless as possible, the table of contents includes page references for both the Larson/Farber text and this manual.

All of the data sets referenced in this manual are found on the data disk packaged in the back of every new copy of Larson/Farber *Elementary Statistics: Picturing the World.* If needed, the MINITAB files (.mtp) may also be downloaded from the texts' companion website at www.prenhall.com/Larson.

▶ Contents:

Getting Started with MINITAB

▶ Using MINITAB Files

MINITAB is a Windows-based Statistical software package. It is very easy to use, and can perform many statistical analyses. When you first open MINITAB, the screen is divided into two parts. The top half is called the Session Window. The results of the statistical analyses are often displayed in the Session Window. The bottom half of the screen is the Data Window. It is called a Worksheet and will contain the data.

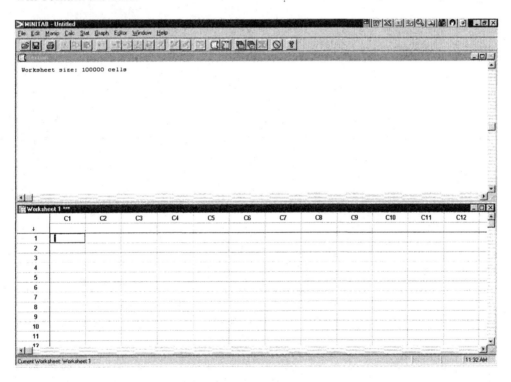

The data can either be entered directly into the Worksheet, or saved worksheets can be opened and used.

▶ Entering Data into the Data Window

To enter the data into the Data Window, you must first click on the bottom half of the screen to make the Data Window active. You can tell which half of the screen is active by the blue bar going across the screen. In the previous picture, notice that the blue bar is in the middle of the screen, highlighting **Worksheet 1.** This indicates that the Data Window is active. The bar will be gray if the Window is not active. (Notice the Session Window bar is gray.)

In MINITAB, the columns are referred to as C1, C2, etc. Notice that there is an empty cell directly below each heading C1, C2, etc. This cell is for a column name. Column names are optional because you can refer to a column as C1 or C2, but a name helps to describe the data contained in a column. Enter the data beginning in cell 1. Notice that the cell numbers are located in the leftmost column of the worksheet.

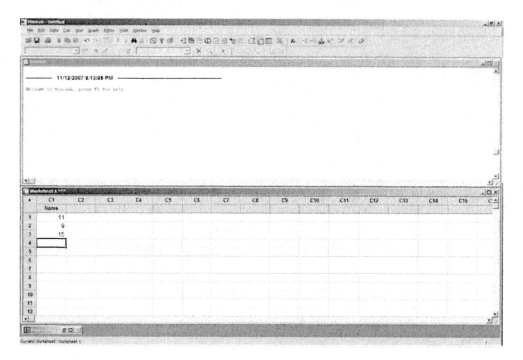

▶ Opening Saved Worksheets

Many of the worksheets that you will be using are saved on the enclosed data disk. To open a saved worksheet, click on **File → Open Worksheet.** The following screen will appear.

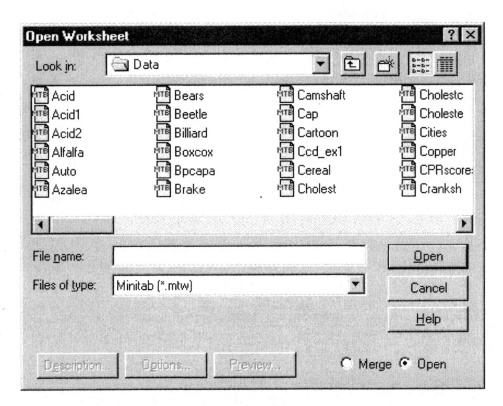

First, you must tell MINITAB where the data files are located. Since the data files are located on the data disk, you must tell MINITAB to **Look In** the **Compact Disc (D:)**. To do this, click on the down arrow to the right of the top input field and select your CD drive by double-clicking on it.

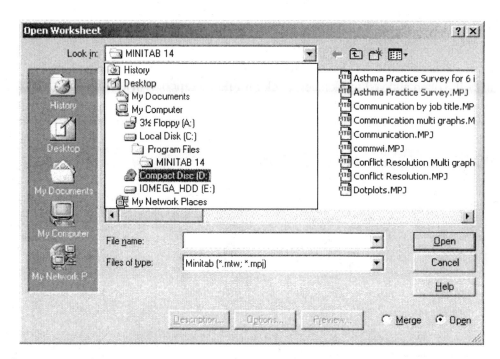

When you do this, you should see three folders listed. Select the MINITAB folder with a double-click. Now you should see a folder for each of the eleven chapters of the book.

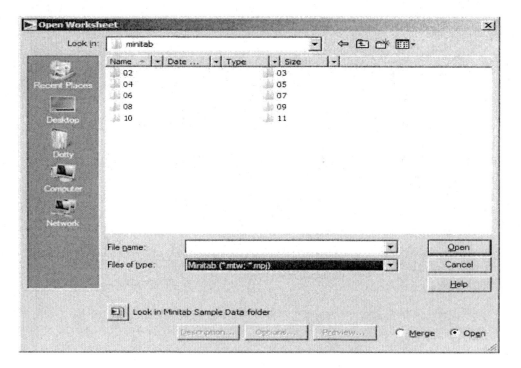

All data files are saved as MINITAB Portable worksheets and have the extension
.mtp. Click on the down arrow for the field called **Files of type** and select
Minitab Portable (*.mtp).

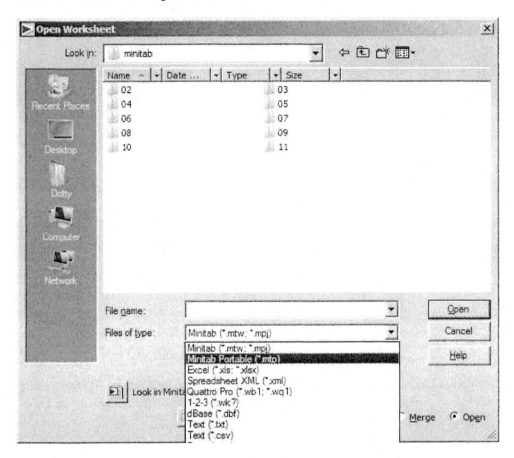

Now, select the folder called **02** (by double-clicking) and you should see all the
MINITAB worksheets for Chapter 2.

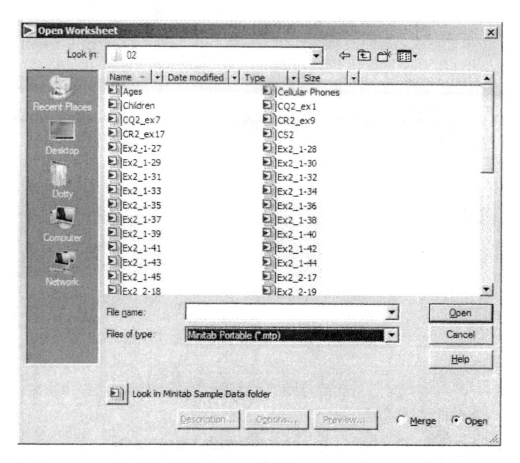

As you can see, the folder **02** has many worksheets saved to disk. To open the
worksheet **Ages**, double-click on it and the worksheet should appear in the Data
Window.

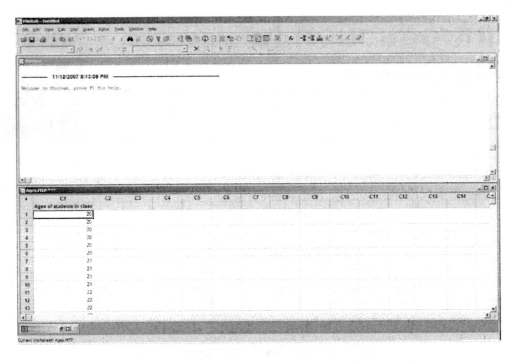

You are now ready to begin analyzing the data and learning more about MINITAB.

Introduction to Statistics

CHAPTER

1

▶ Technology Exercises (pg. 37) Generating Random Numbers

1. To select 10 randomly selected brokers from the company's 86, click on
    Calc → Random Data → Integer. The **Number of rows of data to
    generate** is 10, and then **Store in column(s)** C1. The broker numbers will
    have a **Minimum value** 1 and **Maximum value** 86.

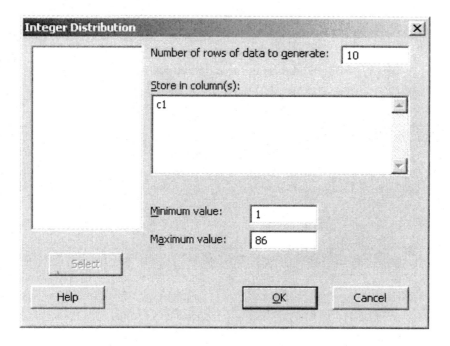

Click on **OK** and the 10 random numbers should be in C1 of the Data
Window.

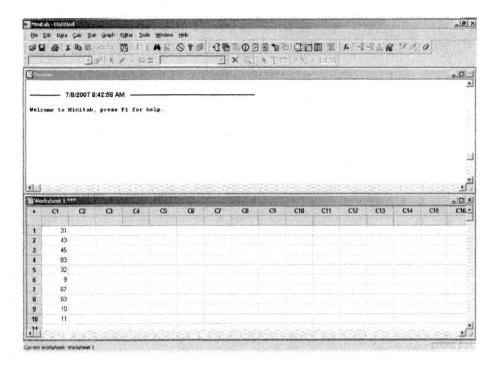

To order the sample list, click on **Data → Sort.** You should **Sort column:**
C1, **By column:** C1 and **Store sorted data in → Original Column**(s).

Click on **OK** and C1 should contain the sorted sample of 10 brokers.

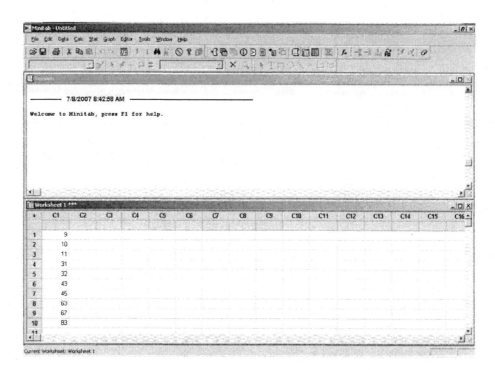

Since this is a *random* sample, each student will have different numbers in C1.

2. For this problem, use the same steps as above. Click on **Calc → Random Data → Integer.** The **Number of rows of data to generate** is 25, and then **Store in column(s)** C1. The camera phone numbers will have a **Minimum value** 1 and **Maximum value** 300. Finally, to order the sample list, click on **Data → Sort.** You should **Sort column:** C1, **By column:** C1 and **Store sorted data in → Original Column(s).**

3. First store the integers 0 to 9 in C1 since you will want to be able to calculate an average later. Click on **Calc → Make Patterned Data → Simple Set of Numbers.** You should **Store patterned data in** C1. The numbers will begin **From the first value** 0 and go **To last value** 9 **In steps of** 1. Next, to randomly sample 5 digits, click on **Calc → Random Data → Sample from columns.** You need to **Sample** 5 **rows from column** C1 and **Store the sample in** C2. Repeat this two more times and **store the sample in** C3 and then in C4. Now you have generated the three samples.

Now, to find the average of each of the four columns (C1, C2, C3, and C4), click on **Calc → Column Statistics.** The Statistic that you would like to calculate is the mean, so click on **Mean.** Enter C1 for the **Input variable** and click on **OK.**

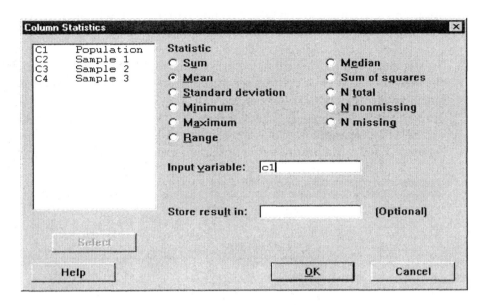

The population mean will be displayed in the Session Window. Repeat this for C2, C3, and C4.

The following output will appear in the Session Window; however, since this is random data, your means will be different for Columns 2 - 4.

Mean of Population
```
Mean of Population = 4.5
```

Mean of Sample 1
```
Mean of Sample 1 = 4.8
```

Mean of Sample 2
```
Mean of Sample 2 = 3.8
```

Mean of Sample 3
```
Mean of Sample 3 = 6.2
```

4. This problem will be very similar to the steps in problem 3. Click on **Calc** → **Make Patterned Data** → **Simple Set of Numbers.** You should **Store patterned data in** C1. The numbers will begin **From the first value** 0 and go **To last value** 40 **In steps of** 1. Next, to randomly sample 7 digits, click on **Calc** → **Random Data** → **Sample from columns.** You need to **Sample 7 rows from column** C1 and **Store the sample in** C2. Repeat this two more times and **store the sample in** C3 and then in C4. Now you have generated the three samples. Now, to find the average of each of the four columns (C1, C2, C3, and C4), click on **Calc** → **Column Statistics.** The Statistic that you would like to calculate is the mean, so click on **Mean.** Enter C1 for the

Input variable and click on **OK.** The population mean will be displayed in the Session Window. Repeat this for C2, C3, and C4.

5. To simulate rolling a 6-sided die, you want to sample with replacement from the integers 1 to 6. You could enter the numbers 1 to 6 into C1 and then sample with replacement. A better way to do this is to click on **Calc →
   Random Data → Integer.** The **Number of rows of data to generate** is 60 and **Store in column** C1. This represents the 60 rolls. Enter a **Minimum value** of 1 and a **Maximum value** of 6 to represent the six sides of the die.

Click on **OK** and C1 should have the results of 60 rolls of a die.

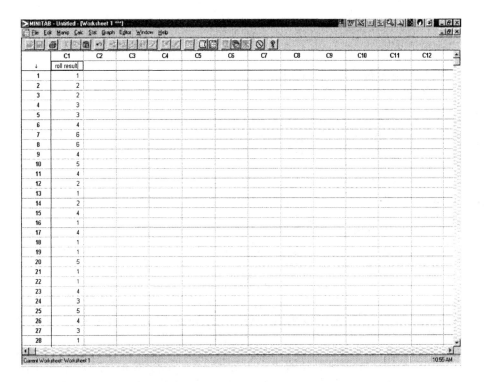

To count how many times each number was rolled, click on **Stat → Tables → Tally Individual Variables.** Enter C1 for the **Variable** and select **Counts** by clicking on it.

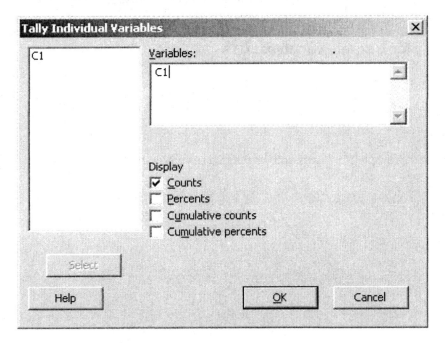

Click on **OK** and the totals will be displayed in the Session Window, as shown below.

Tally for Discrete Variables: C1

C1	Count
1	10
2	12
3	8
4	13
5	9
6	8
N=	60

Recall that each student's results will look different because this is random data. In the results above, 10 ones, 12 twos, 8 threes, 13 fours, 9 fives, and 8 sixes were rolled.

7. To simulate tossing a coin, you can repeat the steps in problem 5. Click on **Calc → Random Data → Integer.** You want to **Generate** 100 **rows of data** and **Store in column** C1. This represents the 100 tosses. Enter a **minimum value** of 0 and a **maximum value** of 1 to represent the two sides of the coin. Click on **OK** and C1 should have the results of 100 tosses of a coin. To count how many times each side of the coined was tossed, click on **Stat → Tables → Tally Individual Variables.** Enter C1 for the **Variable** and select **Counts** by clicking on it. Click on **OK** and the counts will be displayed in the Session Window. Recall that 0 represents heads and 1 represents tails.

Tally for Discrete Variables: C1

C1	Count
0	49
1	51
N=	100

In this example, 49 heads and 51 tails were rolled.

Descriptive Statistics

CHAPTER

2

Section 2.1

▶ Example 7 (pg. 48): Construct a histogram using the Internet
data

To create this histogram, you must open the worksheet called **Internet**. When
you get into MINITAB, click on: **File → Open Worksheet**. On the screen that
appears, **Look In** your Compact Disc (D:) to see the list of available folders.
Double-click on the Folder that is named "**Minitab**" and then select the **ch02**
folder. Next click on the arrow to the right of the **Files of type** field and select
Minitab Portable (mtp). A list of data files should now appear. Use the right
arrow to scroll through the list of files until you see **internet**. Click on this file,
and then click on **Open.**

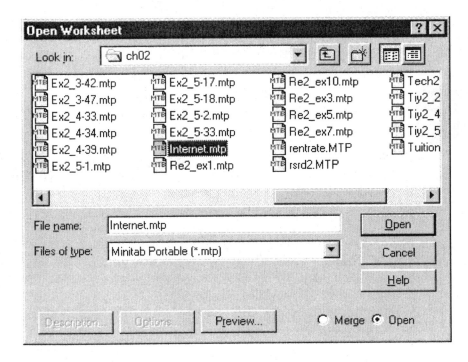

You should now see Internet Subscriber data in Column 1.

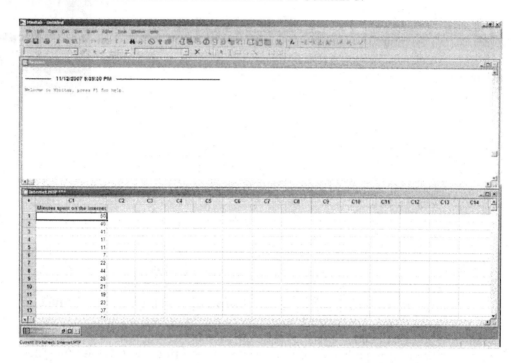

Now you are ready to make the histogram. Click on: **Graph → Histogram**.
Select a **Simple** histogram and click OK.

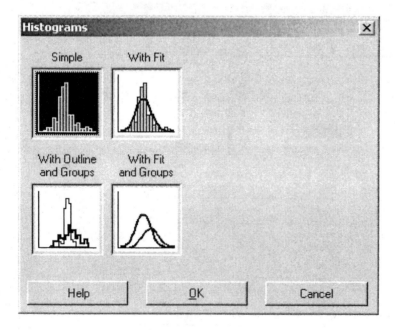

On the main Histogram screen, double-click on C1 in the large box at the left of the screen. "Internet subscribers" should now be filled in as the **Graph variable**.

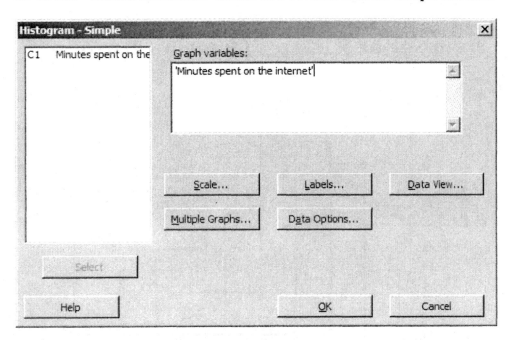

Click on the **Labels** button and enter a title for the graph. Click on **OK**.

At this point, if you click on **OK** again, MINITAB will draw a histogram using default settings. Your histogram will look like the one below.

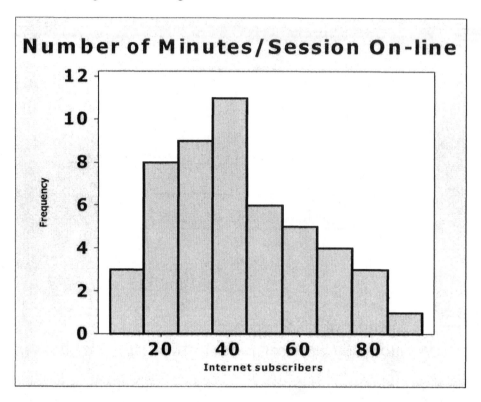

Notice the X-axis label is not "Minutes", and the numbering along the axis is not like the numbering in the textbook. We can edit the graph to fix this, however. Right-click on the X-axis and select **Edit X Scale** from the drop-down menu. Click on the **Binning** tab and check that **Midpoint** is selected. Enter "12.5 : 84.5 / 12" for **Midpoint/Cutpoint Positions**. This tells Minitab you want the numbering to go from 12.5 to 84.5 in steps of 12.

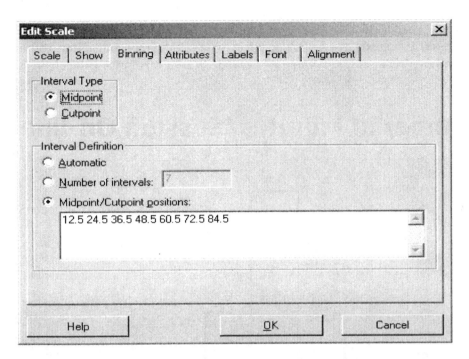

Click on **OK** and the changes should have been made to graph. Next change the X-axis label to "Minutes". Right-click on the current axis label, and select "Edit X axis label" from the drop-down menu. Enter "Minutes" below **Text**.

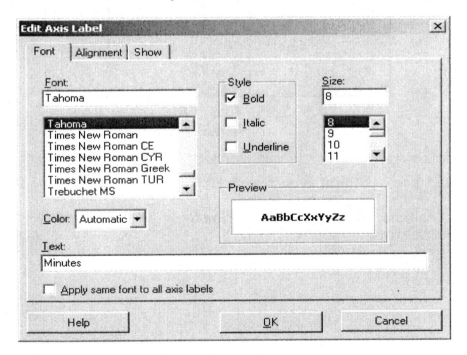

Click on **OK** to view the changes to your histogram.

If you would like to add the frequencies above each rectangle of the histogram (as is shown in the text), right-click on a rectangle of the graph and select **Add → Data Labels.** Click on **Use y-value labels** and click on **OK** to view the changes.

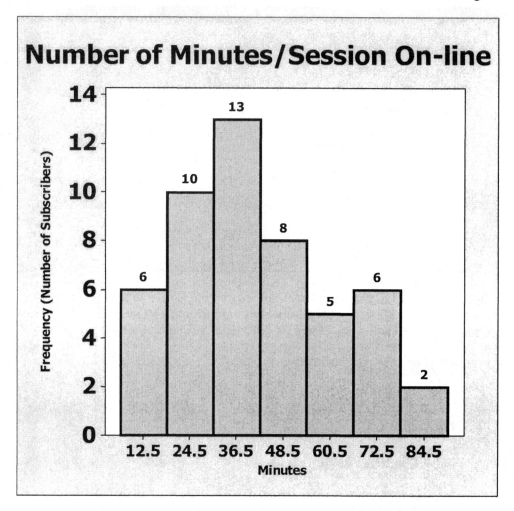

Now your histogram is exactly as in the textbook.

To print the graph, click on **File → Print Graph.** Next, click **OK** and the graph should print.

▶ Exercise 29 (pg. 52) Construct a frequency histogram using 6
Classes

Open the worksheet **ex2_1-29** which is found in the **ch02** MINITAB folder.
Click on: **Graph → Histogram**. Select a **Simple** histogram and click OK.
On the main Histogram screen, double-click on C1 in the large box at the left of
the screen. "July Sales" should now be filled in as the **Graph variable**.

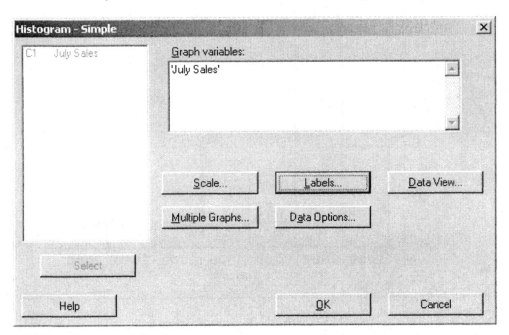

Click on the **Labels** button and enter "July Sales" for the title for the graph.
Click on **OK** twice to view the default histogram that is produced.

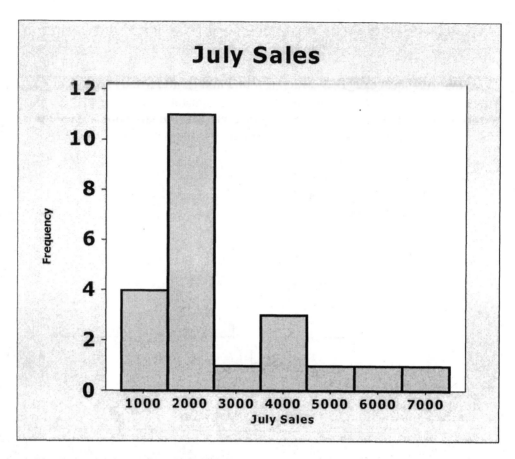

Now change the axis label to "Dollars". To do this, right-click on the axis label
(July Sales), and select "Edit X axis label" from the drop-down menu. Type
"Dollars" below **Text**. Click on **OK** to view the changes.

Next, decide on the numbering for the X-axis that is needed for the 6
classes. This will depend on the class limits you used in your frequency
distribution. In order to use your class limits, you must tell MINITAB to use
Cutpoints. To do this, right-click on the current X-axis numbering. Select "Edit
X Scale" from the drop-down menu. Click on the **Binning** tab. Select **CutPoint**
as the **Interval Type.** Next tell MINITAB what the cutpoint positions are. One
solution is to use cutpoints beginning at 1000 and going up to 7600 in steps of
1100. So fill in 1000:7600/1100 for **Midpoint/cutpoint positions**.

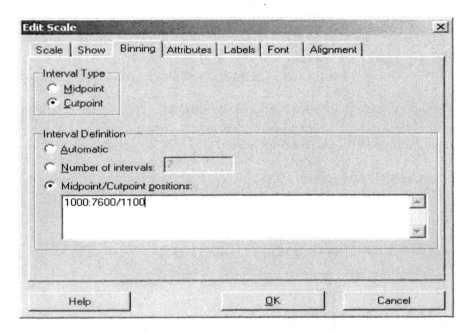

Click on **OK** to view the changes.

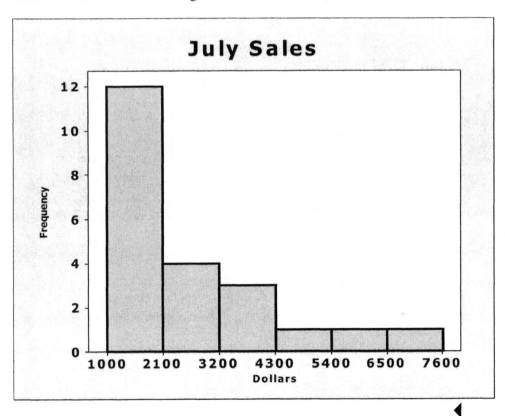

▸ Exercise 33 (pg.52)   Using the bowling scores, construct a
                          relative frequency histogram with 5 classes

Open worksheet **ex2_1-33** which is found in the **ch02** MINITAB folder. Click
on: **Graph → Histogram**. Select a **Simple** histogram and click OK.
On the main Histogram screen, double-click on C1 in the large box at the left of
the screen. "Bowling Score" should now be filled in as the **Graph variable**.
Click on the **Labels** button and enter "Bowling Scores" for the title for the graph.
Click on **OK** twice to view the default histogram that is produced.

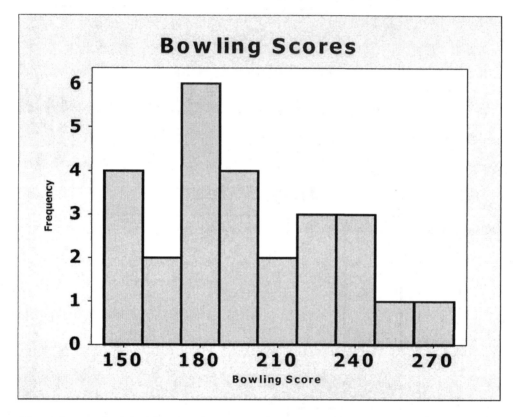

Next, change the label for the X-axis. To do this, right-click on the axis label
(Bowling Scores), and select "Edit X axis label" from the drop-down menu.
Type "Points Scored" below **Text**. Click on **OK** to view the changes.

Now, decide on the numbering for the X-axis that is needed for the 5 classes.
You can either choose the classes yourself as in the last example, or let
MINITAB do it for you. In this example, let MINITAB do it. Right-click on the
X-axis numbers and select "Edit X Scale" from the drop-down menu. Click on

the **Binning** tab. Select **CutPoint** as the **Interval Type.** Next tell MINITAB that the **Number of Intervals** is 5. Click on OK.

Finally, since this is a relative frequency histogram, the right-click on the Y-axis numbering, and select "Edit Y Scale" from the drop-down menu. Click on the **Type** tab and select **Percent.** Click on **OK** twice, and the histogram should have all of the changes.

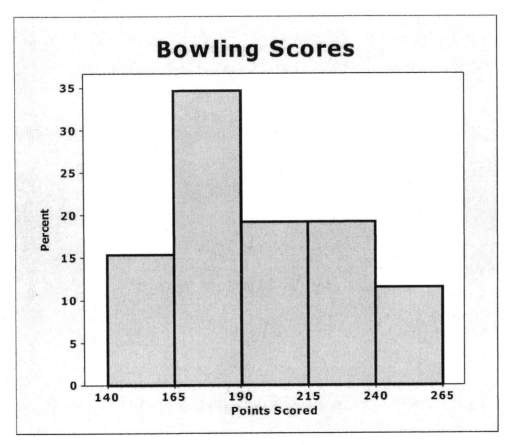

▸ Exercise 34 (pg. 52) Construct a relative frequency histogram
for the ATM data using 5 classes

Open worksheet cx2_1-34 which is found in the ch02 MINITAB folder. Click
on: **Graph** → **Histogram**. Select a **Simple** histogram and click OK.
On the main Histogram screen, double-click on C1 in the large box at the left of
the screen. "ATM withdrawals" should now be filled in as the **Graph variable**.
Click on the **Labels** button and enter "Daily Withdrawals" for the title for the
graph. Click on **OK** twice to view the default histogram that is produced.
Next, change the label for the X-axis. To do this, right-click on the axis label,
and select "Edit X axis label" from the drop-down menu. Type "Hundreds of
Dollars" below **Text**. Click on **OK** to view the changes. Now, decide on the
numbering for the X-axis that is needed for the 5 classes. You can either choose
the classes yourself as in the last example, or let MINITAB do it for you. In this
example, let MINITAB do it. Right-click on the X-axis numbers and select "Edit
X Scale" from the drop-down menu. Click on the **Binning** tab. Select **CutPoint**
as the **Interval Type**. Next tell MINITAB that the **Number of Intervals** is 5.
Click on OK. Finally, since this is a relative frequency histogram, the right-click
on the Y-axis numbering, and select "Edit Y Scale" from the drop-down menu.
Click on the **Type** tab and select **Percent**. Click on **OK** twice, and the histogram
have all of the changes.

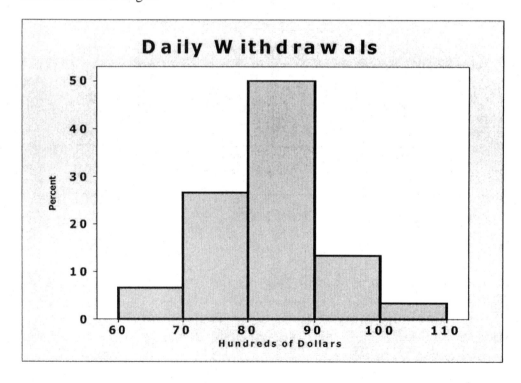

Section 2.2

▶ Example 2 (pg. 56) Constructing a Stem-and-Leaf Plot

Open the file **Text Messages** which is found in the **ch02** MINITAB folder. This worksheet contains the numbers of text messages sent last month by cellular phone users. The data should appear in C1 of your worksheet.

To construct a Stem-and-leaf plot, click on **Graph → Stem-and-Leaf.**
On the screen that appears, select C1 as your **Variable** by doubling clicking on C1. Click on **OK.**

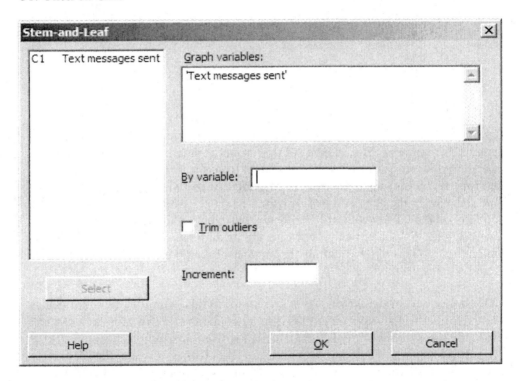

The stem and leaf plot will be displayed in the Session Window.

Stem-and-Leaf Display: Messages

```
Stem-and-leaf of Messages   N  = 50
Leaf Unit = 1.0
   1    7    8
   1    8
   1    8
   1    9
   1    9
   1   10
   6   10    58999
  11   11    22234
  19   11    67888999
  (8)  12    11222344
  23   12    6666699
  16   13    02334
  11   13    99
   9   14    024
   6   14    5578
   2   15
   2   15    59
```

In this MINITAB display, the first column on the left is a counter. This column counts the number of data points starting from the smallest value (at the top of the plot) down to the median. It also counts from the largest data value (at the bottom of the plot) up to the median. Notice that there is only one data point in the first row of the stem and leaf. There are no data points in rows 2 or 6, so the counter on the left remains at "1". Row 7 has 5 data points so the counter increases to "6". The row that contains the median has the number "8" in parentheses. This number counts the number of data points that are in the row that contains the median.

The second column in the display is the **Stem**. In this example, the Stem values range from 7 to 15. Notice that this display contains two rows for each of the values from 8 through 15. These are called **split-stems**. For each stem value, the first row contains all data points with leaf values from 0 to 4 and the second row contains all data points with leaf values from 5 to 9. Notice that MINITAB constructs an *ordered* stem and leaf.

The leaf values are shown to the right of the stem. The leaf values may be the actual data points or they may be the rounded data points. To find the actual values of the data points in the display, use the "Leaf Unit=" statement at the top of the display. The "Leaf Unit" gives you the place value of the leaves. In this stem and leaf, the first data point has a stem value of 7 and a leaf value of 8. Since the "Leaf Unit=1.0", the 8 is the "ones" place and the 7 is in the "tens" places, thus the data point is 78.

▸ Example 3 (pg. 57) Constructing a Dotplot

Open the file **Text Messages** which is found in the **ch02** MINITAB folder. To
construct a dotplot, click on **Graph → Dotplot → Simple.** In the screen that
appears, select C1 for **Graph Variables.** Click on the **Labels** button, enter an
appropriate **Title** and click on **OK.**

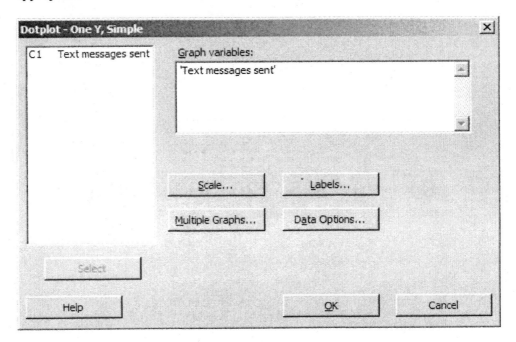

The following dotplot should appear.

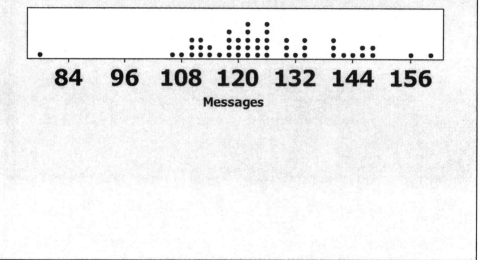

Each dot in the above plot represents the text messages for an individual caller. For example, one caller sent 78 messages, one sent 108, and 3 sent 109 messages (the 3 dots above 109). You can edit the numbering along the X-axis just as in the Histograms.

▶ Example 4 (pg. 58) Constructing a Pie Chart

In this example, you must enter the data into the Data Window. Begin with a clean worksheet. From the table in the left margin of page 58 of the textbook, enter the Vehicle types into C1 and the number of occupants killed into C2. Label each column appropriately as shown below. Note: **OMIT the commas in the number killed.**

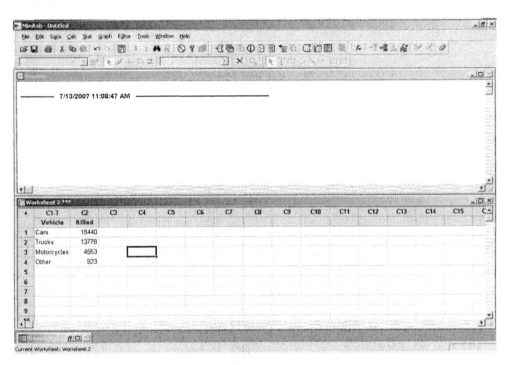

To construct the pie chart, click on **Graph → Pie Chart**. In the screen that appears, select **Chart values from a table**. Enter C1 for the **Categorical Variable** and C2 for the **Summary Variable**. Click on the **Labels** button and enter an appropriate title. Once in the **Labels** screen, you can also select **Slice Labels.** For this example, select **Category Name** and **Percent**.

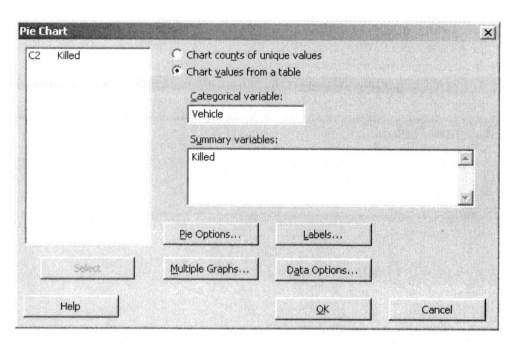

Click on **OK** to view the pie chart.

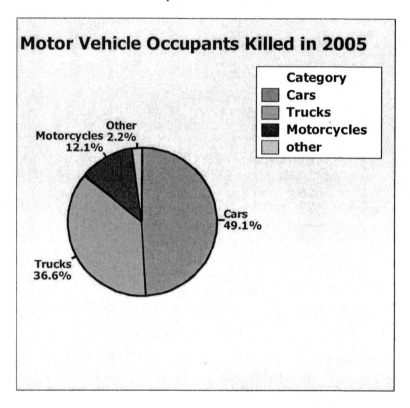

▶ Example 5 (pg. 59) Constructing Pareto Charts

Enter the Inventory Shrinkage data (found in the paragraph for Example 5 on
page 59 of the text) into C1 and C2. Do not include the Total amount of $41.0
million. Note: Do NOT enter the $ signs into C2.

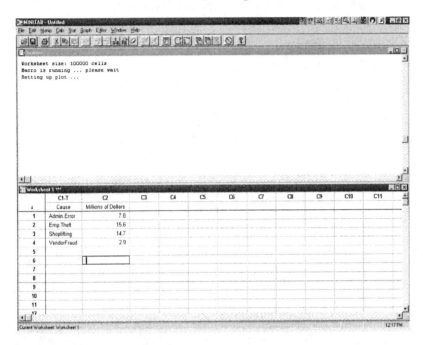

To make the Pareto chart, click on **Graph → Bar Chart**. The **Bars represent**
Values from a table. Select a **Simple** chart and click on **OK**.

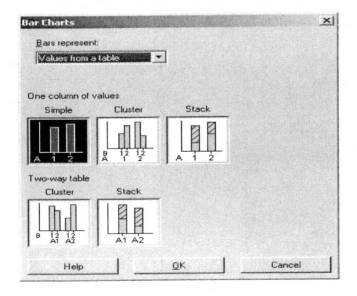

Enter C2 for **Graph variables** and C1 for **Categorical variable.** Click on **Bar chart options** and select **Decreasing Y.** Click on the **Labels** button and enter an appropriate title. Now view the chart.

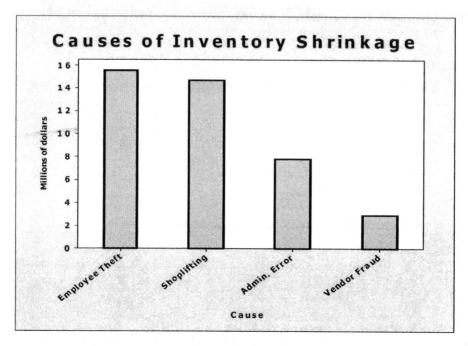

Notice that the default chart has spaces between the rectangles. To remove these spaces, just edit the graph. Right-click on the X-axis and select "Edit X Scale" from the drop-down menu. On the **Scale** tab, click on **Gap between clusters** and enter a 0. Click on **OK** to view the pareto chart.

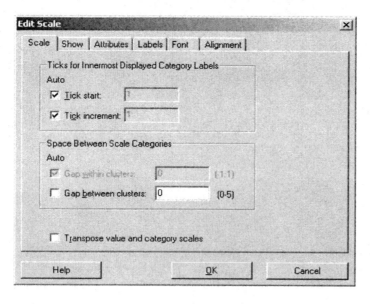

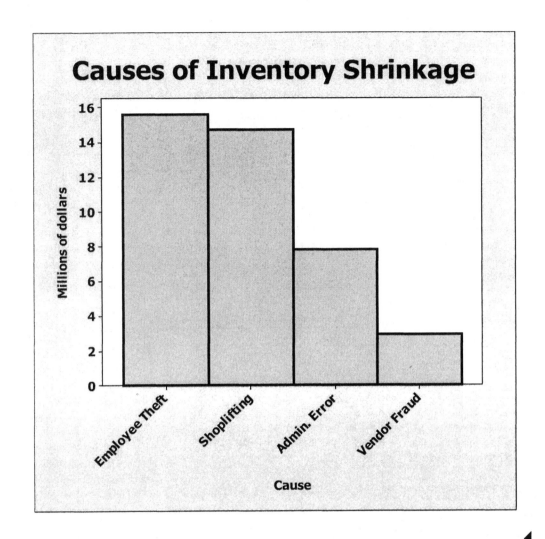

▶ Example 7 (pg. 61) Construct a Time Series Chart of cellular
                      telephone subscribers

Open worksheet **Cellular Phones** which is found in the **ch02** MINITAB folder.
Click on **Graph → Time Series Plot → Simple.** Select C2 as the **Series**.

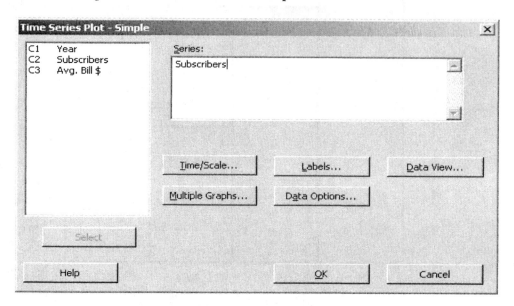

Click on the **Time/Scale** button. Select **Stamp** and enter C1 (Year) for the
Stamp Columns. Click on **OK**.

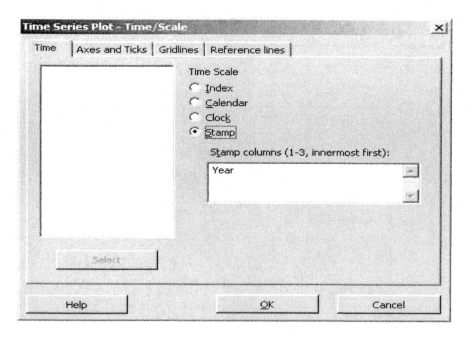

Click on the **Labels** button and enter an appropriate title for the plot.

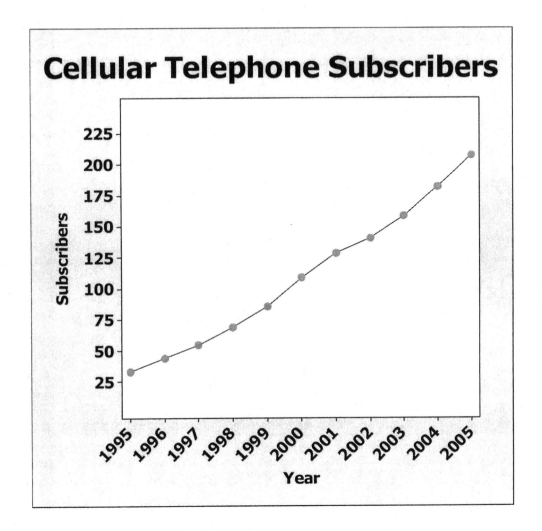

▶ Exercise 17 (pg. 63) Construct a stem and leaf of the exam
scores on a Biology Midterm.

Open worksheet **ex2_2-17** which is found in the **ch02** MINITAB folder. Click
on **Graph→ Stem-and-Leaf.** Select C1 for the **Graph Variable.** Click on **OK**
and the stem and leaf plot should be in the Session Window.

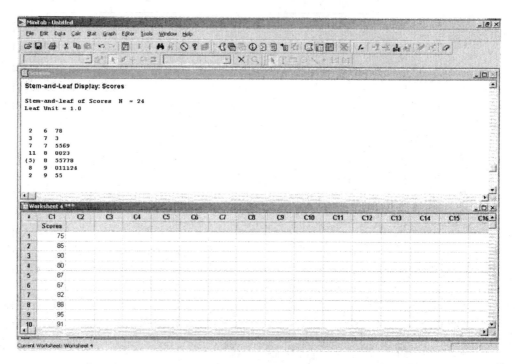

▶ Exercise 22 (pg. 64) Construct a dotplot of the lifespan (in days)
of houseflies

Open worksheet **ex2_2-22** which is found in the **ch02** MINITAB folder. Click
on **Graph → Dotplot → Simple.** Select C1 for the **Graph Variable.** Click on
the **Labels** button and enter an appropriate **Title.** Click on **OK** to view the
dotplot.

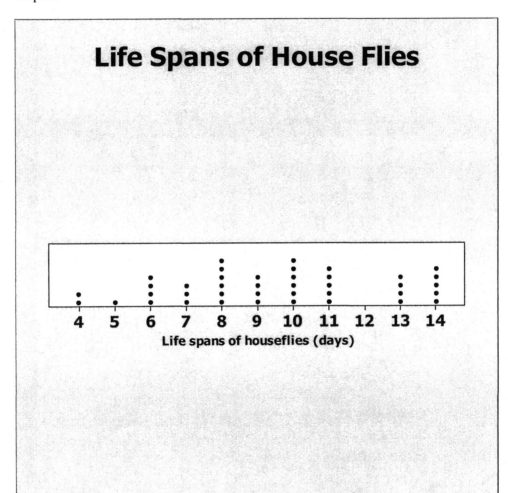

▶ Exercise 24 (pg. 64) Construct a Pie Chart of the data.

This data must be entered into the Data Window. Enter the categories into C1
and the Expenditures into C2. Click on **Graph → Pie Chart.** Select **Chart
values from a table**. Enter C1 for the **Categorical Variable** and C2 for the
Summary Variable. Click on the **Labels** button, enter an appropriate **Title** and
click on **OK**. (You can also select **Slice Labels**.)

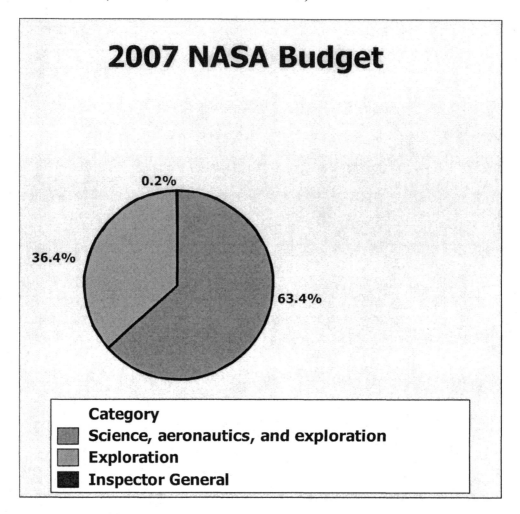

2007 NASA Budget

0.2%

36.4%

63.4%

Category
■ Science, aeronautics, and exploration
■ Exploration
■ Inspector General

▶ Exercise 26 (pg. 64) Construct a Pareto Chart to display the
                           data

Enter the data into the Data Window. Enter the cities into C1 and the ultraviolet
index for each city into C2. To make the Pareto chart, click on **Graph → Bar
Chart**. The **Bars represent** Values from a table. Select a **Simple** chart and
click on **OK**. Enter C2 for **Graph variables** and C1 for **Categorical variable.**
Click on **Bar chart options** and select **Decreasing Y.** Click on the **Labels**
button and enter an appropriate title. Now view the default chart. Notice that the
default chart has spaces between the rectangles. To remove these spaces, just
edit the graph. Right-click on the X-axis and select "Edit X Scale" from the
drop-down menu. On the **Scale** tab, click on **Gap between clusters** and enter a
0. Click on **OK** to view the pareto chart.

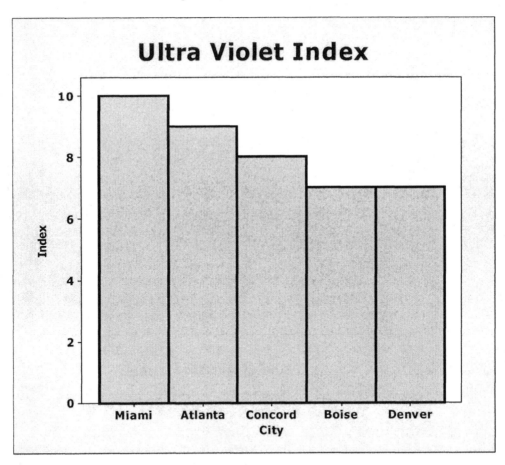

▸ Exercise 28 (pg. 65) Construct a Scatterplot of the data.

Enter the Number of Students per Teacher into C1 and the Average Teacher's Salary into C2 (data found to the left of exercise). Click on **Graph** → **Scatterplot** → **Simple.** Select C2 for the **Y variable** and C1 for the **X variable.** Click on the **Labels** button and enter an appropriate title. Click on **OK** twice to view the scatterplot.

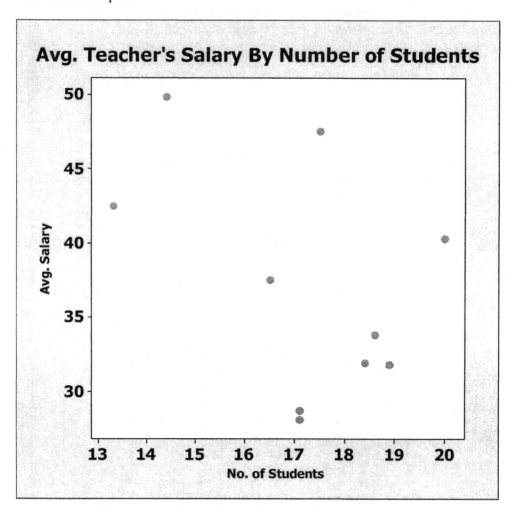

▸ Exercise 31 (pg. 65) Construct a Time Series Plot of the price of
eggs.

Enter the data into the Data Window. Enter the Years into C1 and the Price of
Eggs into C2. Click on **Graph → Time Series Plot → Simple.** Select C2 as the
Series. Click on the **Time/Scale** button. Select **Stamp** and enter C1 (Year) for
the **Stamp Columns.** Click on **OK**. Click on the **Labels** button and enter an
appropriate title for the plot. Click on **OK** to view the plot.

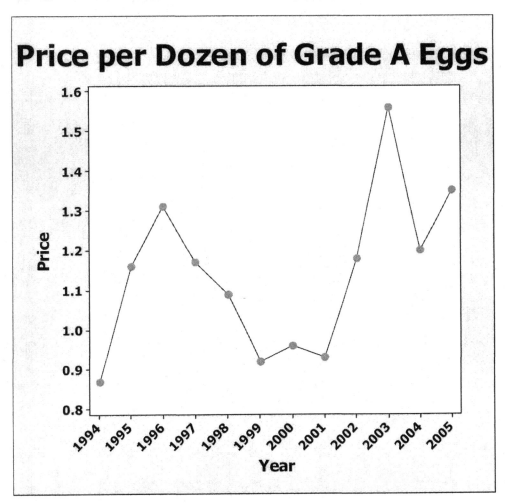

Section 2.3

> **Example 6 (pg. 70)** Find the mean and standard deviation of the
> age of students

Finding the mean and standard deviation of a dataset is very easy using
MINITAB. Open the worksheet **Ages** which is found in the **ch02** MINITAB
folder. Click on **Stat → Basic Statistics → Display Descriptive Statistics.** You
should see the input screen below.

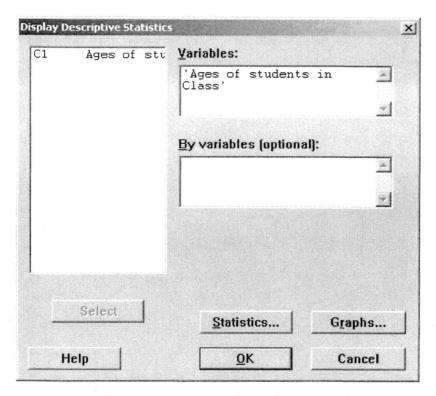

Double click on C1 to select the age data that is entered in C1. Click on **OK** and
the descriptive statistics should appear in the Session Window.

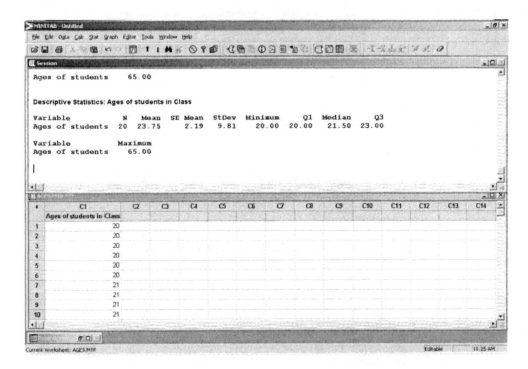

Notice that MINITAB displays several descriptive statistics: sample size, mean, standard error of the mean, standard deviation, standard error of the mean, minimum value, maximum value, median, and the first and third quartiles.

The mode is NOT produced by the above procedure, however, it is quite simple to have MINITAB tally up the data values for you, and then you can select the one with the highest count. Click on **Stat → Tables → Tally Individual Variables**. On the input screen, double-click on C1 to select it. Also, click on **Counts** to have MINITAB count up the frequencies for you.

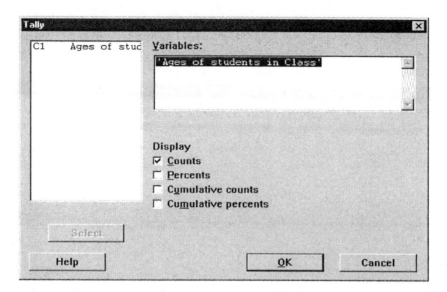

When you click on **OK**, a frequency table will appear in the Session Window.

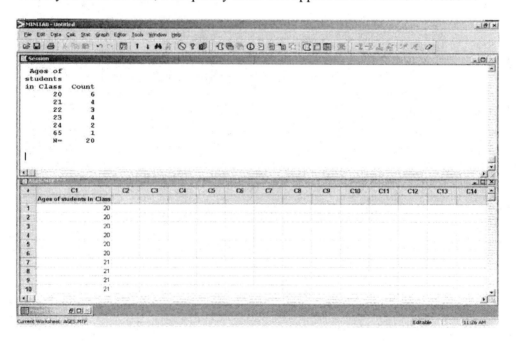

Notice that Age 20 has a count of 6. This means that 6 people in the data were age 20. Since this is the highest count, 20 is the mode.

To print the Session Window with both the descriptive statistics and the frequency table in it, click anywhere up in the Session Window to be sure that it is the active window. Next click on **File → Print Session Window**.

▶ Exercise 21 (pg. 75) Find the mean, median, mode for points per
game scored by each NFL team

Open worksheet **ex2_3-21** which is found in the **ch02** MINITAB folder. Click
on **Stat→ Basic Statistics → Display Descriptive Statistics.** Double-click on
C1 to select it. Click on **OK** and the results should be in the Session Window.
Next, make the frequency table to help find the mode. Click on **Stat → Tables
→ Tally Individual Variables**. Double-click on C1, then click on **OK**. Now
both of the displays will be in the Session Window.

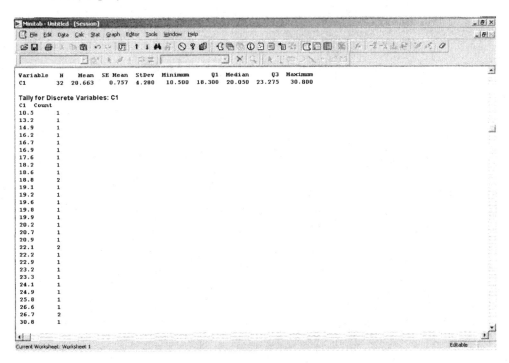

The mode is the points per game with the highest count. In this problem, there is
a 3-way tie for the mode: 18.8, 22.1, and 26.7. Each of these has a count of 2.

▶ Exercise 53 (pg. 79) Construct a frequency histogram of the heights using 5 classes

Open worksheet **ex2_3-53** which is found in the **ch02** MINITAB folder. Click on **Graph → Histogram → Simple.** Select C1 for the **Graph variable.** Click on the **Labels** button and enter an appropriate title. Click on OK to view the default histogram. Since you need 5 classes, edit the graph. Right-click on the X-axis and select "Edit X scale" from the drop-down menu. On the **Binning** tab, select **Cutpoints,** and enter 5 for the **Number of Intervals.** Click on **OK** to view the changes to the graph.

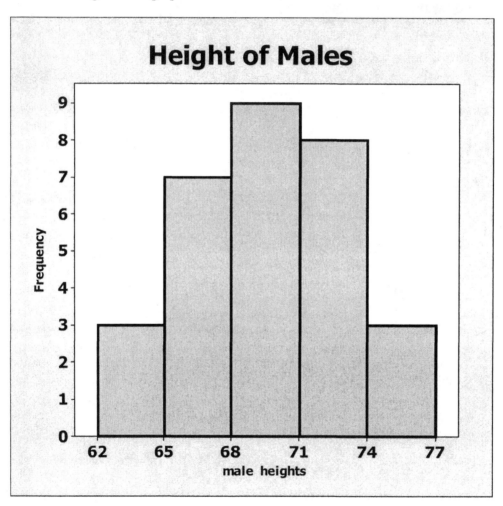

▶ Exercise 60 (pg. 80) Find the mean, median, and construct a
stem and leaf plot of the data

Open worksheet ex**2_3-60** which is found in the **ch02** MINITAB folder. Click
on **Stat → Basic Statistics → Display Descriptive Statistics.** Select C1 for the
Variable and click on **OK.** The descriptive statistics should be in the Session
Window. Next, click on **Graph → Stem-and-Leaf.** Select C1 for the **Variable**
and click on **OK.** The stem and leaf plot should also be in the Session Window
as shown in the next picture.

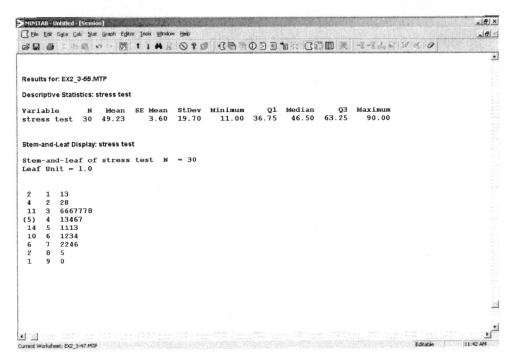

Section 2.4

▶ Example 5 (pg. 86) Calculate the mean and standard deviation for the office rental rates in Miami

Open worksheet **rentrate** which is found in the **ch02** MINITAB folder. Click on **Stat → Basic Statistics → Display Descriptive Statistics.** Select C1 for the **Variable** and click on **OK.** The descriptive statistics should be in the Session Window.

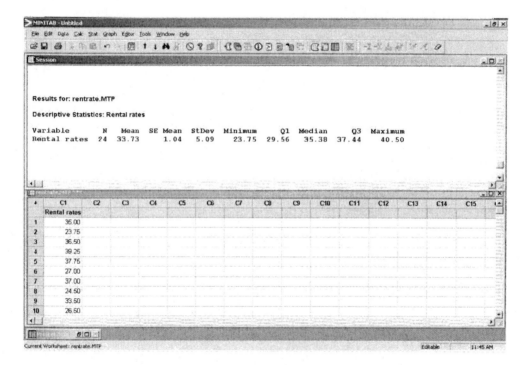

▶ Exercise 2 (pg. 92) Find the range, mean, variance, standard
deviation of the dataset.

Enter the data into C1 of the Data Window. Click on **Stat → Basic Statistics →
Display Descriptive Statistics.** Select C1 for the **Variable** and click on **OK.**
The descriptive statistics should be in the Session Window.

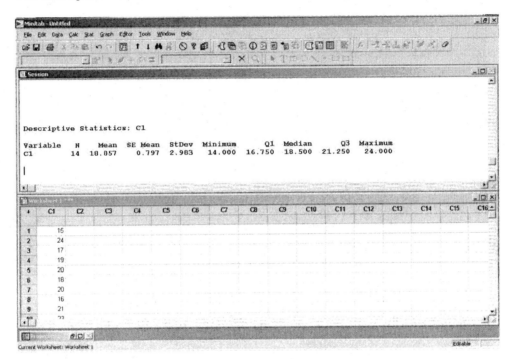

▶ Exercise 19 (pg. 94) Find the mean, range, standard deviation
and variance for each city

Enter the data into C1 and C2 of the Data Window. Be sure to label the columns.
Click on **Stat → Basic Statistics → Display Descriptive Statistics.** Select both
C1 and C2 for the **Variable** and click on **OK.** The descriptive statistics should
be in the Session Window.

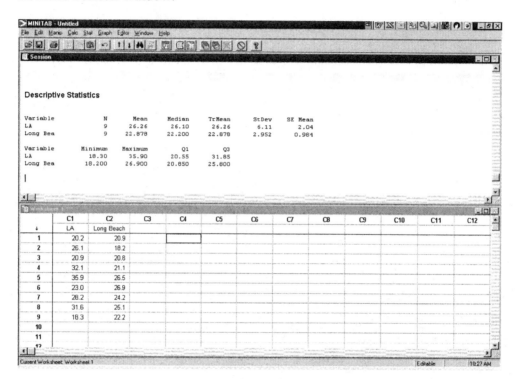

Section 2.5

▶ Example 2 (pg. 103) Find the first, second, and third quartiles of the tuition data.

Open worksheet **Tuition** which is found in the **ch02** MINITAB folder. Click on **Stat → Basic Statistics → Display Descriptive Statistics.** Select C1 for the **Variable** and click on **OK.** The descriptive statistics should be in the Session Window. Recall that the median is the second quartile.

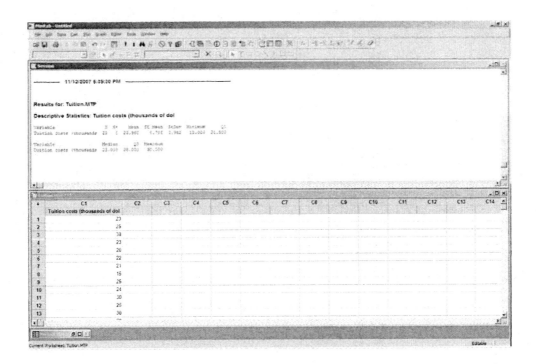

▶ Example 4 (pg. 105) Construct a box-and-whisker plot using the test scores given in Example 1.

Enter the data found on page 93 of the text into C1 of the Data Window. Click on **Graph → Boxplot → Simple**. Select C1 for the **Graph variable.** Next, since by default MINITAB plots vertically, click on the **Scale** button and select **Transpose value and category scales.** This will turn the plot horizontal, as in the textbook. Click on the **Labels** button and enter an appropriate title. Click on **OK** twice to view the boxplot.

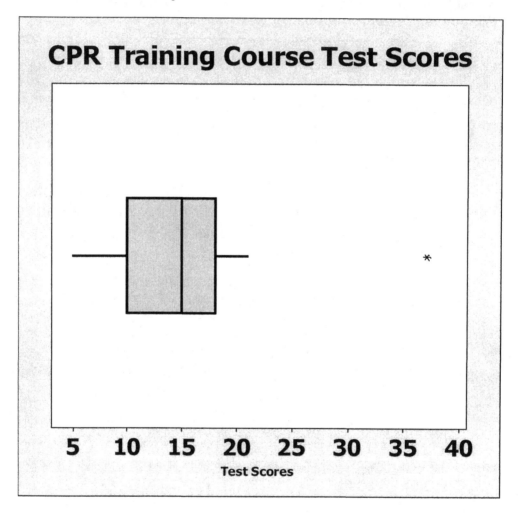

▶ Exercise 2 (pg. 109) Draw a boxplot of the data

Open worksheet **ex2_5-2** which is found in the **ch02** MINITAB folder. Click on
Graph → Boxplot → Simple. Select C1 for the **Graph variable.** Next, since
by default MINITAB plots vertically, click on the **Scale** button and select
Transpose value and category scales. This will turn the plot horizontal, as in
the textbook. Click on the **Labels** button and enter an appropriate title. Click on
OK twice to view the boxplot.

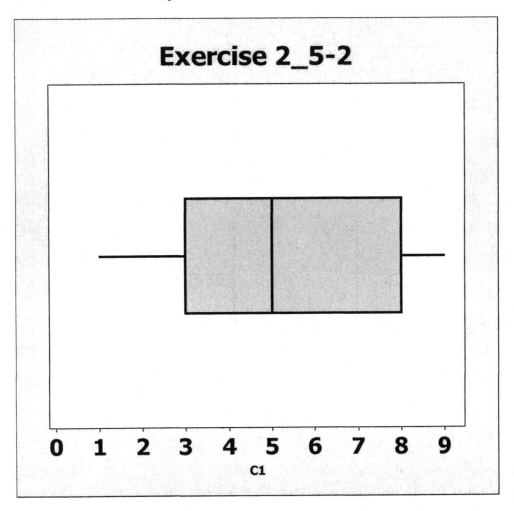

> Exercise 23 (pg. 110) Draw a boxplot of hours of TV
>   watched per day

Open worksheet **ex2_5-23** which is found in the **ch02** MINITAB folder. Click
on **Graph** → **Boxplot** → **Simple**. Select C1 for the **Graph variable.** Next,
since by default MINITAB plots vertically, click on the **Scale** button and select
Transpose value and category scales. This will turn the plot horizontal, as in
the textbook. Click on the **Labels** button and enter an appropriate title. Click on
OK twice to view the boxplot.

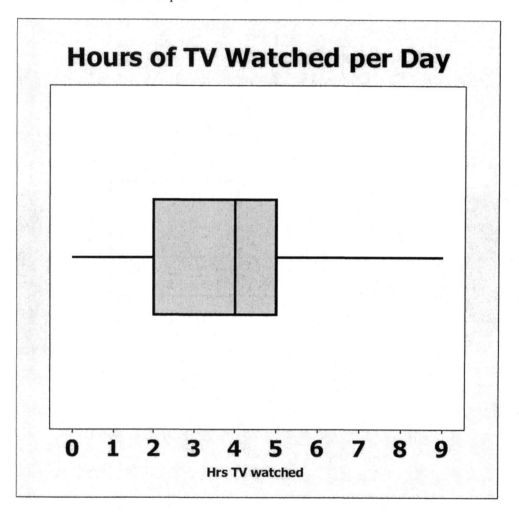

▶ Technology Lab (pg. 123) Use descriptive statistics and a
histogram to describe the milk data

Open worksheet **Tech2** which is found in the **ch02** MINITAB folder. Click on
Stat → Basic Statistics → Display Descriptive Statistics. Click on **OK** and the
descriptive statistics will be displayed in your Session Window.

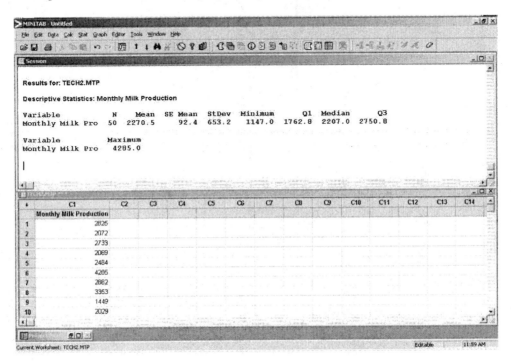

Next construct the histogram. Click on **Graph → Histogram → Simple.** Select
C1 for the **Graph variable.** Click on the **Labels** button and enter an appropriate
title. Click on OK to view the default histogram. Since you need a class width
of 500, edit the graph. Right-click on the X-axis and select "Edit X scale" from
the drop-down menu. On the **Binning** tab, select **Cutpoints.** Using the
information from the descriptive statistics, you can see that the minimum value is
1147 and the maximum is 4285. Since you want a class width of 500, use
positions beginning at 1100 and going up to 4600 in steps of 500. Thus, on the
Binning tab enter the **Midpoint/cutpoint positions** as **1100:4300/500.** Click on
OK to view the changes to the graph.

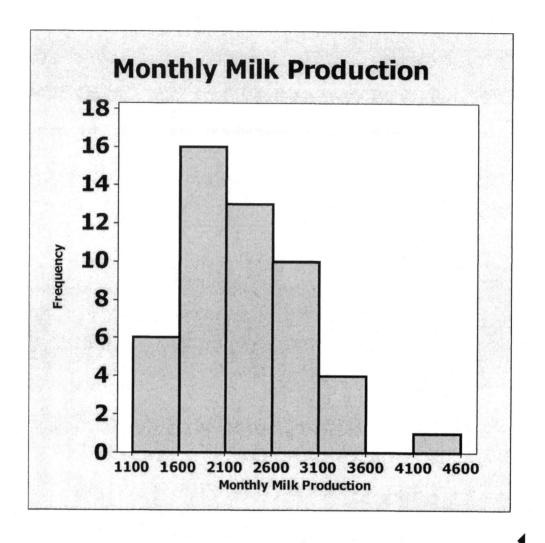

Probability

Section 3.1

▸ Law of Large Numbers (pg. 138) Coin Simulation

You can use MINITAB to simulate repeatedly tossing a fair coin and then calculate the empirical probability of tossing a head. This empirical probability will more closely approximate the theoretical probability as the number of tosses gets large. To do this simulation, generate 1000 "tosses" of a fair coin. Let "0" represent a head, and "1" represent a tail. Click on **Calc → Random Data → Bernoulli**. **Generate** 1000 **rows of data** and **Store in column** C1. Use .5 for the **Probability of Success.** When you click on **OK**, you should see C1 filled with 1's and 0's. To count up the number of "0"s, click on **Stat → Tables → Tally Individual Variables.** Select C1 for **Variables** (by double clicking on C1), and choose both **counts and percents** by clicking on the box to the left of each one. When you click on **OK**, the summary statistics will appear in the Session Window. In the following example, notice that there were 511 heads and 489 tails. Thus, the empirical of tossing a head is .511 (51.1%). This is a very good approximation of the theoretical probability of .5.

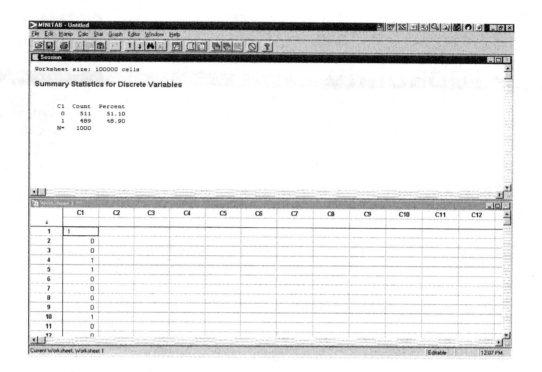

Section 3.2

> **Exercise 36d (pg. 159)** Birthday Problem

Simulate the "Birthday Problem" using MINITAB. To do this simulation, the days of the year will be represented by the numbers 1 to 365. Click on **Calc → Random Data → Integer. Generate** 24 **rows of data** (representing the 24 people in the room), **Store In** C1, **Minimum value** is 1 (representing Jan.1st) and **Maximum value** is 365 (representing Dec. 31st). When you click on **OK**, you should see C1 filled with 24 numbers ranging between 1 and 365. These numbers represent the birthdays of the 24 students in the class. The question is: Are there at least two people in the room with the same birthday? To answer this question, we must first summarize the data from this simulation. To do this, click on **Stat → Tables → Tally Individual variables.** On the screen that appears, select C1 for **Variables** and select **Counts**. Next click on **OK**, and in the Session Window, you will see a summary table of C1. This table lists each of the different birthdays that occurred in this simulation, as well as a count of the number of people who had that birthday. Notice that most counts are 1's. If you see a count of "2" or more, then you have at least two people in the room with the same birthday.

Repeat the simulation 9 more times and tally the results each time. How many of the 10 columns had at least two people with the same birthday? The empirical probability that at least two people in a room of 24 people will share a birthday can be calculated as follows: (# of columns having at least two people with the same birthday) divided by 10 (which is the total number of simulations).

▶ Technology Lab (pg. 191)   Composing Mozart Variations

3. Click on **Calc → Random Data → Integer. Generate 1 row of data, Store In Column** C1, **Minimum value** is 1 and **Maximum value** is 11. For Part B, repeat these steps but **Generate** 100 **rows of data** (instead of 1 row). To tally the results, click on **Stat → Tables → Tally Individual variables.** Select both **Counts** and **Percents.** The results will appear in the Session Window. Compare the percents to the theoretical probabilities you found in Part A.

5. Click on **Calc → Random Data → Integer. Generate 2 rows of data, Store In** C1, **Minimum value** is 1 and **Maximum value** is 6. Add the two numbers and subtract 1 to obtain the total. For Part B, **Generate** 100 **rows of data, Store In Columns** C1-C2, **Minimum value** is 1 and **Maximum value** is 6. The total will be calculated for each row by adding the two numbers from C1 and C2 and then subtracting 1. To do this, click on **Calc → Calculator. Store result in variable** C3. For **Expression,** type in the following: C1 + C2 - 1. Click on **OK,** and the totals should be in C3. To tally the results, click on **Stat → Tables → Tally Individual variables.** Select both **Counts** and **Percents.** The results will appear in the Session Window. Compare the percents to the theoretical probabilities you found in Part A.

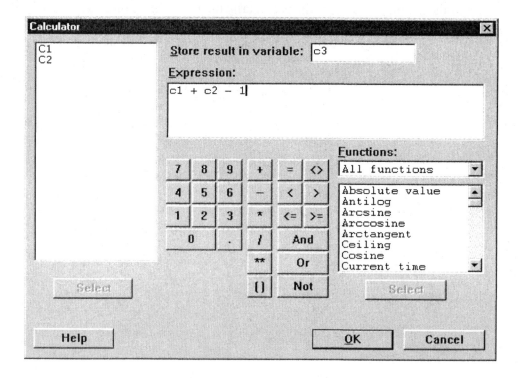

To choose a minuet, Mozart suggested that the player toss a pair of dice 16 times. For the 8th and 16th bars, choose Option 1 if the dice total is odd, and Option 2 if the dice total is even. For each of the other 14 bars, subtract 1 from the dice total. To do this in MINITAB, first simulate rolling the dice 16 times. Click on **Calc → Random Data → Integer. Generate** 16 **rows of data, Store In Columns** C1-C2, **Minimum value** is 1 and **Maximum value** is 6. The total will be calculated for each row by adding the two numbers from C1 and C2 and then subtracting 1. To do this, click on **Calc → Calculator. Store result in variable** C3. For **Expression** type in: C1+C2-1. Click on **OK**, and the totals should be in C3.

	C1	C2	C3	C4	C5	C6	C7	C8	C9	C10	C11	C12
	Roll 1	Roll 2	Total-1									
1	3	6	8									
2	3	4	6									
3	1	1	1									
4	3	4	6									
5	2	4	5									
6	5	6	10									
7	6	1	6									
8	1	2	2									
9	5	4	8									
10	6	5	10									
11	3	1	3									
12	6	6	11									
13	6	4	9									
14	6	1	6									
15	5	1	5									
16	6	1	6									
17												
18												
19												
20												
21												
22												
23												
24												
25												
26												
27												
28												

The numbers in C3 will be the minuet, except for the 8th and 16th bars. To find these, add C1 + C2 for rows 8 and 16. If the total is odd, choose Option 1 and if the total is even, choose Option 2. For example, the total in row 8 is 3 and Option 1 should be chosen. The total in row 16 is 7, and Option 1 should be chosen again. Thus, the minuet for this simulation is:

8	6	1	6	5	10	6	1
8	10	3	11	9	6	5	1

Notice the 8th and 16th bars are both 1.

Discrete Probability Distributions

CHAPTER

4

Section 4.2

▶ Example 4 (pg. 210)   Find the probability that 65 out of 100 households own a gas grill

In this example, 59% of American households own a gas grill and a random sample of 100 American households is selected. Thus n = 100 and p = .58. Click on **Calc → Probability Distributions → Binomial.** To find the probability that exactly 65 of the 100 households own a gas grill, select **Probability**. This tells MINITAB what type of calculation you want to do. The **Number of Trials** is 100 and the **Probability of Success** is .59. To find the probability of 65 households owning a gas grill, click on the circle to the left of **Input Constant** and enter 65 in the box to the right of **Input Constant**. Leave all other fields blank. Click on **OK.**

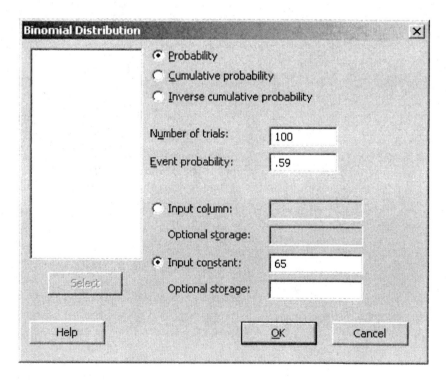

The probability that 65 of the 100 households sampled own a gas grill will be displayed in the Session Window. Notice that the probability is .0391072.

Probability Density Function

```
Binomial with n = 100 and p = 0.59

 x   P( X = x )
65    0.0391072
```

> ▶ **Example 7 (pg. 213)** Graphing a Binomial Distribution

In order to graph the binomial distribution, you must first create the distribution and save it in the Data Window. In C1, type in the values of X. Since n=6, the values of X are 0, 1, 2, 3, 4, 5, and 6. Next, use MINITAB to generate the binomial probabilities for n=6 and p=0.59. Click on **Calc → Probability Distributions → Binomial.** Select **Probability.** The **Number of Trials** is 6 and the **Probability of Success** is .59. Select **Input Column** by clicking on the circle on the left. Now, tell MINITAB that the X values are in C1 and that you want the probabilities stored in C2 by entering C1 as the **Input Column** and entering C2 for **Optional Storage.**

Binomial Distribution	✕

C1

    ⦿ Probability
    ○ Cumulative probability
    ○ Inverse cumulative probability

    Number of trials: `6`

    Event probability: `.59`

    ⦿ Input column: `C1`

        Optional storage: `c2`

    ○ Input constant:

        Optional storage:

[Select]

[Help]   [OK]   [Cancel]

Click on **OK**. The probabilities should now be in C2. Label C1 as "X" and C2 as "P(X)". This will be helpful when you graph the distribution.

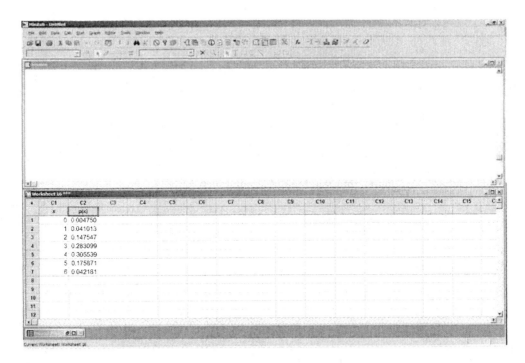

To create the graph, click on **Graph → Bar Chart.** On the screen that appears, below **Bars represent,** click on the down arrow and select **Values from a table.** In the section labeled **One column of values**, select the **Simple** Bar Chart. Click on **OK**. On the screen that appears, select 'C2 P(X)' for the **Graph variable** and select 'C1 X' for the **Categorical variable**. Click on **Labels** and enter an appropriate title. Click on **OK** twice to view the graph.

On the graph, right click on the label ('X') for the X-axis. Click on **Edit X label** and, in the box for **Text**, enter 'Households.' Next, right click on any numerical value on the X-axis (for example, right click on '1' under the first bar of the chart.) Click on **Edit X scale**. On the screen that appears, click on the check mark to the left of **Gap between clusters**. (This will remove the check mark.) In the box to the right of **Gap between clusters**, enter '0'. Click on OK and the new graph will appear.

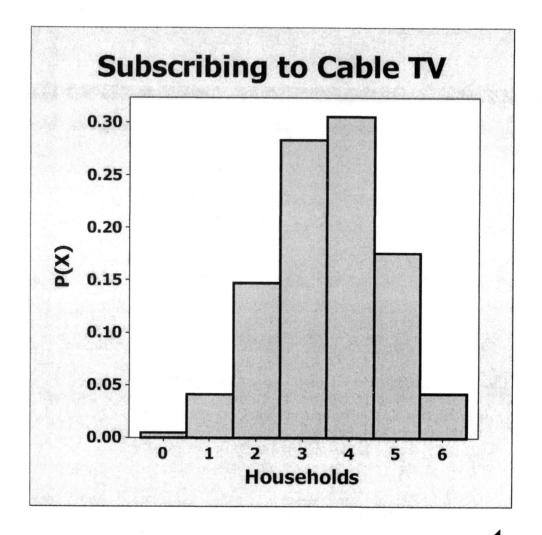

Section 4.3

▶ Example 3 (pg. 224) Finding Poisson Probabilities

Since there is an average of 3.6 rabbits per acre living in a field, $\mu = 3.6$ for this Poisson example. To find the probability that 2 rabbits are found on any given acre of the field, click on **Calc → Probability Distributions → Poisson.** Since you want a simple probability, select **Probability** and enter 3.6 for the **Mean.** To find the probability that X=2, select **Input constant** and enter 2 for the value.

Poisson Distribution	✕
⦿ **P**robability	
○ **C**umulative probability	
○ **I**nverse cumulative probability	
Mean: `3.6`	
○ Input co**l**umn: `___`	
Optional s**t**orage: `___`	
⦿ Input co**n**stant: `2`	
Optional sto**r**age: `___`	
Select	
Help	**O**K Cancel

Click on **OK** and the probability will be displayed in the Session Window.

Probability Density Function

```
Poisson with mean = 3.6

x   P( X = x )
2     0.177058
```

▶ Technology Lab (pg. 237)

1. First, create the Poisson distribution and save it in the Data Window. In C1, type in the values of X. Since n=20, the values of X are 0, 1, 2, 3, 4, 5, ...20. Next, use MINITAB to generate the Poisson probabilities for n=20 and μ=4. Click on **Calc → Probability Distributions → Poisson.** Select **Probability**. The **Mean** is 4. Now, tell MINITAB that the X values are in C1 and that you want the probabilities stored in C2 by entering C1 as the **Input Column** and entering C2 for **Optional Storage.** Click **OK.** The probabilities will be displayed in C2 of the Data Window. Notice, for example, that P(X=4) = .195367. This probability is the height at X=4 on the histogram that is displayed in the upper right corner of pg. 237.

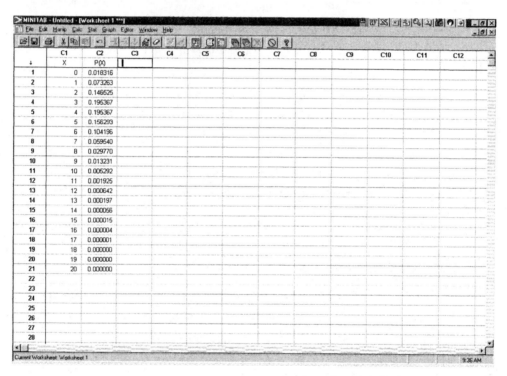

3. To generate 20 random numbers from a Poisson distribution with mean=4, click on **Calc → Random Data → Poisson. Generate** 20 **rows of data** and **Store in column** C3. Enter a **Mean** of 4 and click on **OK.**

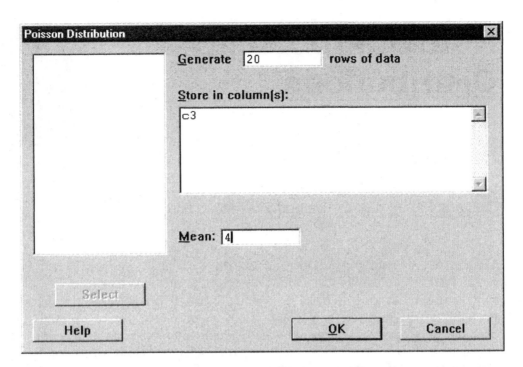

Use the random numbers that are in C3 of the Data Window to create the table of waiting customers and the probabilities. Click on **Stat → Tables → Tally Individual Values.** The **Variable(s)** is C3. Select **Percents**, and click on **OK.**

5. Repeat the steps in Exercise 3, but enter a **Mean** of 5 this time and **Store in column** C4.

6. To calculate P(X=10) for a Poisson random variable with a mean of 5, click on **Calc → Probability Distributions → Poisson.** Since you want a simple probability, select **Probability** and enter 5 for the **Mean** and 10 for the **Input constant.**

7. To find the probabilities for parts a - c, use the Poisson probability distribution that you created in C1 and C2.

Normal Probability Distributions

CHAPTER

5

Section 5.2

▶ Example 3 (pg. 255) Using MINITAB to find Normal Probabilities.

Cholesterol levels of American men are normally distributed with μ=215 and σ=25. Find the probability that a randomly selected American man has a cholesterol level that is less than 175. To do this in MINITAB, click on **Calc** → **Probability Distributions** → **Normal.** On the input screen, select **Cumulative probability.** (Cumulative probability *'accumulates'* all probability to the left of the input constant.) Enter 215 for the **Mean** and 25 for the **Standard deviation.** Next select **Input Constant** and enter the value 175.

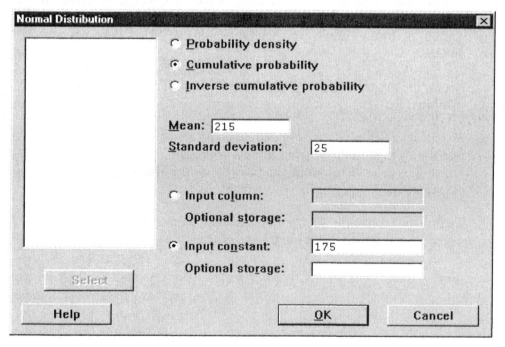

Click on **OK** and the probability should be displayed in the Session Window. As you can see, the probability that a randomly selected American man has a cholesterol level that is less than 175 is equal to .0547993.

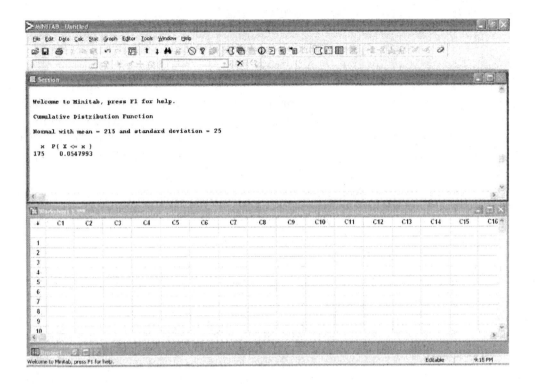

▶ Exercise 13 (pg. 257) Height of American Males

Heights are normally distributed with μ=69.6 inches and σ=3.0 inches. To complete parts (a) (c), you will need MINITAB to give you two probabilities: one using X=66 and the other using X=72. Click on **Calc → Probability Distributions → Normal.** On the input screen, select **Cumulative probability.** Enter 69.2 for the **Mean** and 3.0 for the **Standard deviation.** Next select **Input Constant** and enter the value 66. Click on **OK.** Repeat the above steps using an **Input constant** of 72. Now the Session Window should contain $P(X \leq 66)$ and $P(X \leq 72)$

Cumulative Distribution Function

```
Normal with mean = 69.6 and standard deviation = 3

  x   P( X <= x )
66      0.115070
```

Cumulative Distribution Function

```
Normal with mean = 69.6 and standard deviation = 3

  x   P( X <= x )
72      0.788145
```

So, for Part (a), the $P(X \leq 66)$ = .115070. For Part (b), to find the $P(66 \leq X \leq 72)$, you must subtract the two probabilities. Thus, the $P(66 \leq X \leq 72)$ = .788145 - .115070 = .673075. For Part (c), to find $P(X > 72)$, it is 1 - .824676 = .211855.

◀

Section 5.3

▶ **Example 4 (pg. 264)** Finding a specific data value

Scores for a civil service exam are normally distributed with μ=75 and σ=6.5. To be eligible for employment, you must score in the top 5%. Find the lowest score you can earn and still be eligible for employment. To do this in MINITAB, click on **Calc → Probability Distributions → Normal.** On the input screen, select **Inverse Cumulative probability.** Enter 75 for the **Mean** and 6.5 for the **Standard deviation.** For this type of problem, the **Input constant** will be the area to the left of the X-value we are looking for. This input constant will be a decimal number between 0 and 1. For this example, select **Input Constant** and enter the value .95 since 5% of the test scores are above this number and therefore, 95% are below this number. Click on **OK** and the X-value should be in the Session Window. Notice that the test score that qualifies you for employment is 85.6915.

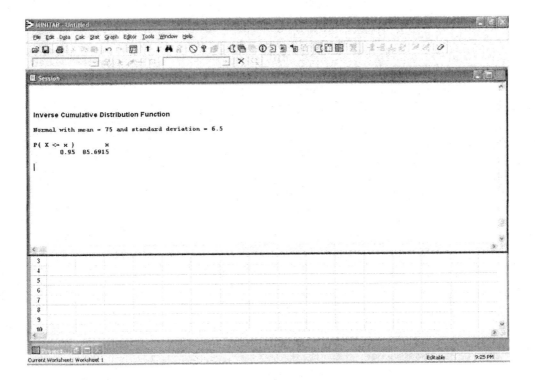

Section 5.4

▶ Example 6 (pg. 277) Finding Probabilities for X and $\overline{X}$

Credit card balances are normally distributed, with a mean of $2870 and a standard deviation of $900.

1. To find the probability that a randomly selected card holder has a balance less than $2500, click on **Calc → Probability Distributions → Normal.** On the input screen, select **Cumulative probability.** Enter 2870 for the **Mean** and 900 for the **Standard deviation.** Next select **Input Constant** and enter the value 2500. Click on **OK** and the probability should appear in the Session Window.

2. To find the probability that the *mean* balance of 25 card holders is less than $2500, you will need to calculate the standard deviation of $\overline{x}$ which is equal to $900/\sqrt{25} = 180$. (Use a hand calculator for this calculation.) Now let MINITAB do the rest for you. Click on **Calc → Probability Distributions → Normal.** On the input screen, select **Cumulative probability.** Enter 2870 for the **Mean** and 180 for the **Standard deviation.** Next select **Input Constant** and enter the value 2500. Click on **OK** and the probability should appear in the Session Window.

◀

▶ Technology Lab (pg. 303) Age Distribution in the United States

In this lab, you will compare the age distribution in the United States to the sampling distribution that is created by taking 36 random samples of size n=40 from the population and calculating the sample means.

Open worksheet **Tech5** which is found in the **ch05** folder. C1 should now contain the mean ages from the 36 random samples.

1. From the table on page 303 in your textbook, enter the Class Midpoints into C2. To enter the midpoints, click on **Calc → Make Patterned Data → Simple Set of Numbers.** On the input screen, you should **Store patterned data in** C2, **from the first value** of 2 **to the last value** of 97, **in steps of** 5. Click on **OK** and the midpoints should now be in C2. Next enter the relative frequencies, converted to proportions, into C3. So for a relative frequency of 6.8%, you should enter .068 into C3. The mean of this distribution is Σx p(x). To calculate the mean, you will have to multiply C2 and C3. To do this, click on **Calc → Calculator.** Type in the **Expression** C2 * C3 and **store result in variable** C4. Click on **OK** and in C4 you should now see the product of C2 and C3.

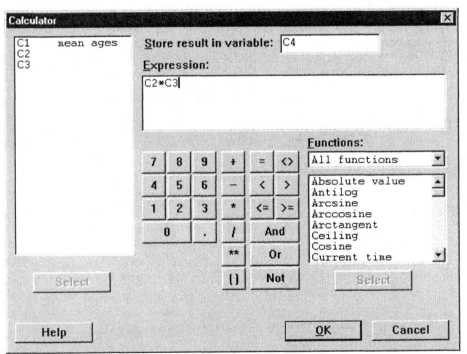

Now find the sum of C4. Click on **Calc → Column Statistics.** On the input screen, select **Sum** and use C4 for the **Input variable.** Click on **OK** and the column sum should be in the Session Window. As you can see, the mean age in the United States is 36.59.

2. The 36 sample means are in C1. To find the mean, click on **Stat → Basic Statistics → Display Descriptive Statistics.** Select C1 for the **Variable** and click on **OK**. The descriptive statistics will be displayed in the Session Window. The mean of the set of sample means is 36.209 and the standard deviation is 3.552. (You will need the standard deviation for question 6.)

4. To draw the histogram, click on **Graph → Histogram.** Select C1 for the **Graph variable.** In order to create a *relative frequency* histogram, click on **Options** and select **Percent.** Click on **OK** twice and you should be able to view the histogram.

5. To find the standard deviation of the ages of Americans, you must use the formula for the standard deviation of a Discrete Random variable, found on page 199 in the textbook. The shortcut formula will make this calculation easier. Use the formula $\Sigma x^2 \, p(x) - \mu^2$ and take the square root of this value. In MINITAB, first square all the midpoints. Click on **Calc → Calculator.** Type in the **Expression** C2 * C2 and **store result in variable** C5. Click on **OK** and in C5 you should now see the midpoints squared. To calculate x^2 $p(x)$, you must multiply C5 by C3. Click on **Calc → Calculator.** Type in the **Expression** C5 * C3 and **store result in variable** C6.

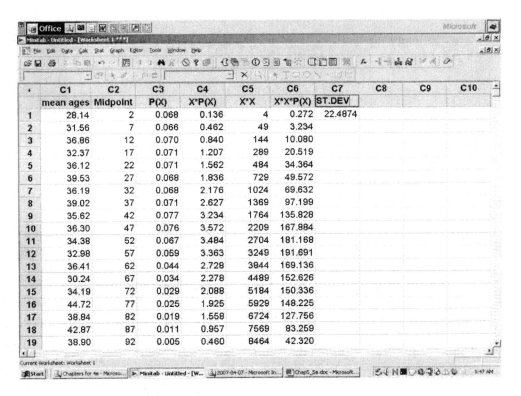

Now find the sum of C6. Click on **Calc → Column Statistics.** On the input screen, select **Sum** and use C6 for the **Input variable.** Click on **OK** and the column sum should be in the Session Window. As you can see, $\Sigma x^2 p(x)$ is 1844.51. Next, subtract μ^2 from 1844.51 and take the square root. (Recall that $\mu = 36.59$) Click on **Calc → Calculator.** Type in the **Expression** SQRT(1844.51-36.59*36.59) and **store result in variable** C7. The standard deviation is the number now in C7, 22.4874.

6. The standard deviation of the 36 sample means can be found in the descriptive statistics that you produced for question 2.

Confidence Intervals

CHAPTER

6

Section 6.1

▶ Example 4 (pg. 314)   Construct a 99% Confidence Interval for the mean number of sentences in an ad

Open the worksheet **Sentences** which is found in the **ch06** Folder. The data will be in C1. First find the standard deviation of the data. Click on **Calc → Column Statistics.** Select **Standard deviation** and enter C1 for the **Input variable.** Click on **OK** and the standard deviation will be displayed in the Session Window. To construct the confidence interval, click on **Stat → Basic Statistics → 1-Sample Z.** Enter C1 in **Samples in columns**. For **Standard deviation,** enter the assumed standard deviation of 5.

1-Sample Z (Test and Confidence Interval)	✕

○ Samples in columns:

    Sentences

○ Summarized data

    Sample size: [   ]

    Mean: [   ]

Standard deviation: [5]

☐ Perform hypothesis test

    Hypothesized mean: [   ]

| Select | | Graphs... | Options... |
| Help | | OK | Cancel |

Next, click on **Options** and enter 99.0 for the **Confidence Level.**

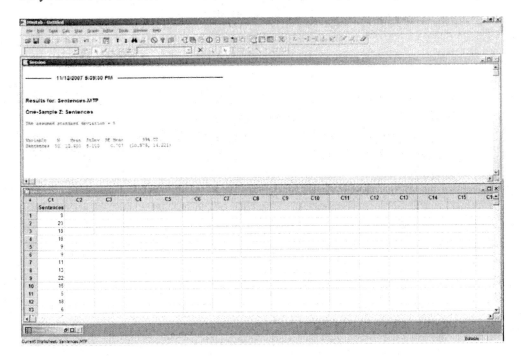

Click on **OK** twice and the results will be displayed in the Session Window.
As you can see, the 99% confidence interval is (10.579, 14.221).

▶ Exercise 51 (pg. 319) Construct 90% and 99% confidence intervals for mean reading time

Enter the data into C1. To construct the confidence interval, click on **Stat →**
Basic Statistics → 1-Sample Z. Enter C1 for **Samples in columns**. Enter the
assumed value of 1.5 for the **Standard deviation**. Next, select **Options** and
enter 90.0 for the **Confidence Level.** Click on **OK** and the interval will be
displayed in the Session Window. Repeat the above steps using 99.0 for the
Confidence Level.

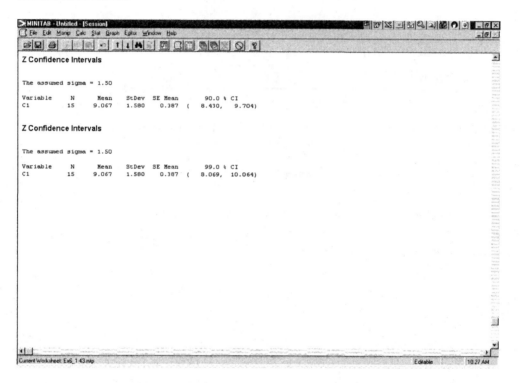

▶ Exercise 65 (pg. 322)   Construct a 95% confidence interval for
airfare prices

Open worksheet **Ex6_1-65.mtp** in the **ch06** Folder. The airfares are in C1. Find
the standard deviation of the sample. Click on **Calc → Column Statistics.**
Select **Standard deviation** and enter C1 for the **Input variable.** Click on **OK**
and the standard deviation will be displayed in the Session Window. To
construct the confidence interval, click on **Stat → Basic Statistics → 1-Sample
Z.** Enter C1 in **Samples in columns**. For **Standard deviation,** enter the standard
deviation that is displayed in the Session Window. Next, select **Options** and
enter 95.0 for the **Confidence Level.** Click **OK** twice.

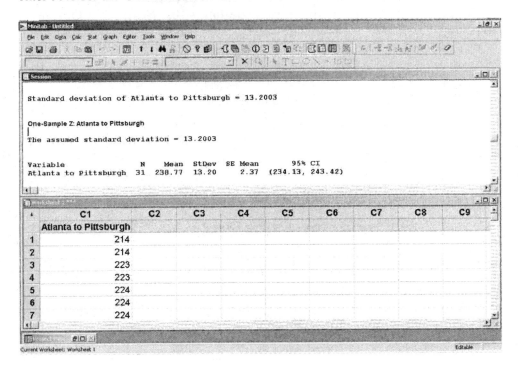

Section 6.2

> ▶ **Example 2 (pg. 327)** Construct a 95% Confidence Interval for the mean temperature of coffee sold

Using the summarized data found on page 327 of the text, construct a 95% confidence interval for the temperatures of the coffee sold at 16 randomly selected restaurants. Since n=16 and the population standard deviation is unknown, you should construct a t-interval for this problem. Click on **Stat →** **Basic Statistics → 1-Sample t.** Select **Summarized data**. The **Sample size** is 16, the **Mean** is 162, and the **Standard deviation** is 10. Next, select **Options** and enter 95.0 for the **Confidence Level.** Click on **OK** twice and the output will be displayed in the Session Window.

1-Sample t (Test and Confidence Interval)

○ **Samples in columns:**

◉ **Summarized data**

Sample size: 16

Mean: 162

Standard deviation: 10

Test mean: _____ (required for test)

Select Graphs... Options...

Help OK Cancel

The following confidence interval will be displayed in the Session Window.

One-Sample T

N	Mean	StDev	SE Mean	95% CI
16	162.000	10.000	2.500	(156.671, 167.329)

▸ **Exercise 21 (pg. 331)**   Construct a 99% confidence interval for
the mean SAT scores

Enter the SAT scores into C1. Since n=12 and the population standard deviation
is unknown, you should construct a t-interval for this problem. Click on **Stat →
Basic Statistics → 1-Sample t.** Enter C1 in **Samples in column**. Next, select
Options and enter 99.0 for the **Confidence Level.** Click on **OK** twice and the
output will be displayed in the Session Window.

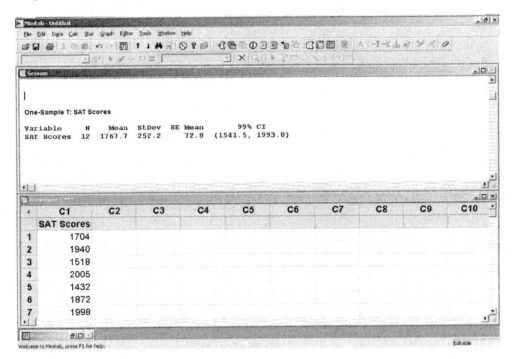

Section 6.3

▶ Example 2 (pg. 336) Construct a 95% confidence interval for p

From Example 1, on page 334 of the textbook, you know that 354 of 1219 American adults said that their favorite sport was football. To construct a 95% confidence interval, click on **Stat → Basic Statistics → 1 Proportion.** Select **Summarized Data.** The **Number of trials** is 1219 and the **Number of events** is 354.

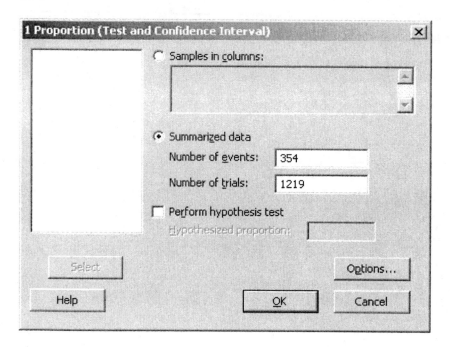

Next, to select the confidence level, click on **Options.** Enter 95.0 for the **Confidence Level.** Also, select **Use test and interval based on normal distribution.**

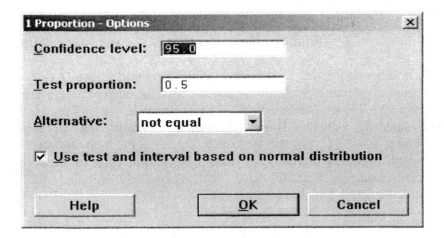

Click on **OK** twice and the output will be displayed in the Session Window.

Test and CI for One Proportion

```
Sample      X      N  Sample p         95% CI
1          354   1219  0.290402   (0.265040, 0.316779)
```

Notice the interval is (.265, .317). This means that with 95% confidence, you can say that the proportion of adults who say football is their favorite sport is between 26.5% and 31.7%.

▶ **Exercise 19 (pg. 340)** Construct intervals for the proportion of
children planning to join a volunteer group

A study of 848 children found that 144 planned to join the armed forces in the
future. Construct both 95% and 99% confidence intervals for the true proportion
of children who plan to enlist. Click on **Stat → Basic Statistics →
1 Proportion.** Select **Summarized Data.** The **Number of trials** is 848 and the
Number of event is 144. Next, to select the confidence level, click on **Options.**
Enter 95.0 for the **Confidence Level.** Also, select **Use test and interval based
on normal distribution.** Click on **OK** twice and the results will be in the
Session Window. Repeat using 99.0 for the **Confidence Level.**

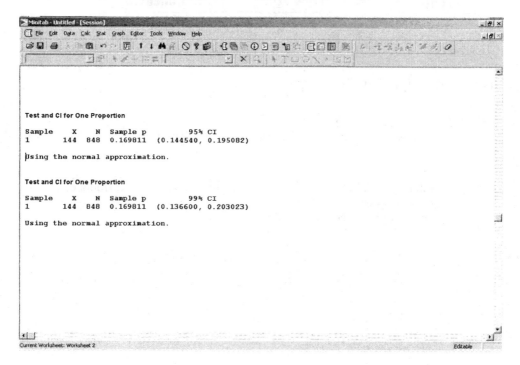

▶ Technology Lab (pg. 359)

1. Click on **Stat → Basic Statistics → 1 Proportion.** Select **Summarized Data.** The **Number of trials** is 1010 and the **Number of events** is 131 (.13 x 1010=131.1). Next, to select the confidence level, click on **Options.** Enter 95.0 for the **Confidence Level.** Also, select **Use test and interval based on normal distribution.** Click on **OK** twice and the results will be in the Session Window.

3. Oprah Winfrey was named by 9% of the people in the sample. That means that 91 (.09 x 1010 = 90.9) of the 1010 named Oprah Winfrey. Use the steps in question 1 to construct the 95% confidence interval. This time the **Number of events** is 91.

4. To do this simulation, you will generate random binomial data with n=1010 and p=.12. The result displayed in each cell of C1 will represent the number of people who named Oprah. Click on **Calc → Random Data → Binomial. Generate** 200 **rows of data** and **Store in column** C1. The **Number of trials** is 1010 and the **Probability of success** is .12. When you click on **OK**, C1 will contain a simulation of 200 samples.

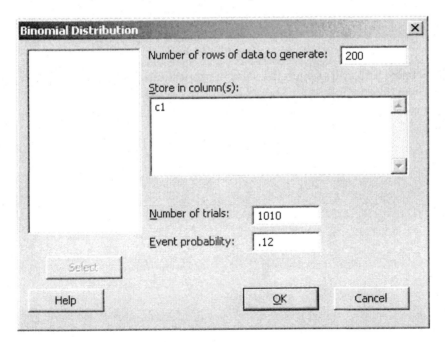

To calculate the sample proportions, click on **Calc → Calculator**. Enter C1/1010 for the **Expression** and **Store the result in** C2. Click on **OK.** Now C2 contains 200 values of the sample proportion. Sort C2 to find the smallest and largest value. Click on **Data → Sort. Sort column** C2, **By column** C2 and **Store sorted data in column of current worksheet** C3.

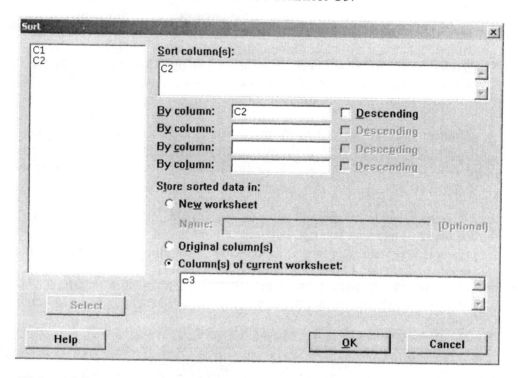

Click on **OK** and C3 should contain the sample proportions sorted from smallest to largest. Thus, the smallest value should be in Row 1 and the largest value should be in Row 200. Since this is random data, results will vary each time this is repeated.

Hypothesis Testing with One Sample

CHAPTER

7

Section 7.2

▶ Example 5 (pg. 383) Hypothesis Testing Using P-values

You think that the average franchise investment information given in the graph on page 351 of the textbook is incorrect, so you randomly select 30 franchises and determine the mean investment for each is $135,000 with a standard deviation of $30,000. Is there enough evidence to support your claim at $\alpha = .05$? Use the P-value to interpret.

Click on **Stat → Basic Statistics → 1-Sample Z.** Click on **Summarized data.** Enter 30 for **Sample size,** 135000 for **Mean,** 30000 for **Standard deviation**, and 143260 for **Test mean.**

Since the claim is "the mean is different from $143,260", you will perform a two-tailed test. Click on **Options**, and set **Alternative** to "not equal".

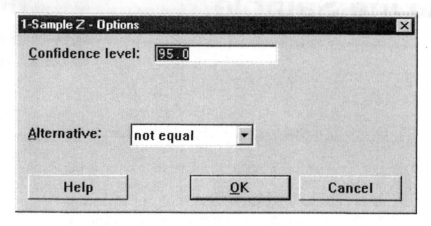

Click on **OK** twice and the results should be displayed in the Session Window.

One-Sample Z

```
Test of mu = 143260 vs not = 143260
The assumed standard deviation = 30000


 N    Mean  SE Mean       99% CI          Z      P
30  135000     5477  (120892, 149108)  -1.51  0.132
```

Notice that both the test statistic and the P-value are given. From the output, note that $z = -1.51$ and $P = .132$. Since the P-value is larger than α, you would fail to reject the null hypothesis.

> ▶ Exercise 37 (pg. 392) Years taken to quit smoking permanently

Open worksheet **Ex7_2-37,** which is found in the **ch07** MINITAB folder.   Click
on **Calc → Column Statistics**. Select **Standard deviation** for the **Statistic** to be
calculated and enter C1 for the **Input Variable.** Click on **OK** and the standard
deviation will be in the Session Window. You will enter this value in the input
screen for the 1-Sample Z test. Click on **Stat → Basic Statistics → 1-Sample Z.**
Click on **Stat → Basic Statistics → 1-Sample Z.** Enter C1 for **Samples in
columns,** 4.288 for **Standard deviation**, and 15 for **Test mean.** Since the claim
is "the mean time is 15 years", you will perform a two-tailed test. Click on
Options and use the down arrow beside **Alternative** to select "not equal". Click
on **OK** and the results of the test should be displayed in the Session Window.

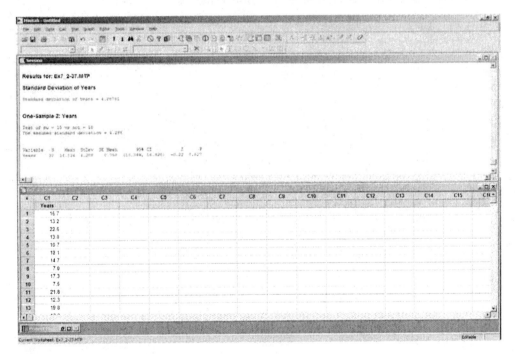

▶ Exercise 43 (pg. 393) Nitrogen Dioxide Level in West London

Open worksheet **Ex7_2-43,** which is found in the **ch07** MINITAB folder. Click on **Calc → Column Statistics.** Select **Standard deviation** for the **Statistic** to be calculated and enter C1 for the **Input Variable.** Click on **OK** and the standard deviation will be in the Session Window. You will enter this value in the input screen for the 1-Sample Z test. Click on **Stat → Basic Statistics → 1-Sample Z.** Enter C1 for **Samples in columns,** 9.164 for **Standard deviation,** and 32 for **Test mean.** Since the claim is "the mean is greater than 32 parts per billion", you will perform a right-tailed test. Click **Options** and then on the down arrow beside **Alternative** and select "greater than". Click on **OK** twice and the results of the test should be displayed in the Session Window.

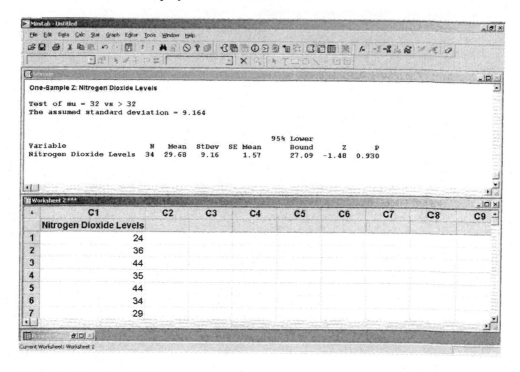

Since the P-value (.93) is greater than the level of significance $\alpha = .05$, the data fail to support the scientist's estimate.

Section 7.3

> ▶ Example 4 (pg. 400) Testing μ with a Small Sample

A used car dealer says that the mean price of a 2005 Honda Pilot LX is at least $23,900. You suspect this claim is incorrect. At $\alpha = .05$, is there enough evidence to reject the dealer's claim?

Based on a random sample of 14 similar vehicles, you found that the mean price is $23,000 with a standard deviation of $1113. Be sure to enter the data in numeric form. So enter the amount $23,000 as 23000. Since this is a small sample problem, you will be performing a 1-Sample t-test. Click on **Stat** → **Basic Statistics** → **1-Sample t.** Click on **Summarized data,** and enter a **Sample size** of 14, a **Mean** of 23000, and a **Standard deviation** of 1113. Click on **Perform hypothesis test** and enter 23900 for **Hypothesized mean**. Since you suspect that the used car dealer's claim is too high, you will perform a left-tailed test. Click on **Options** and then on the down arrow beside **Alternative** to select "less than". Click on **OK** and the results of the test should be displayed in the Session Window.

One-Sample T

Test of mu = 23900 vs < 23900

| | | | | 95% Upper | | |
N	Mean	StDev	SE Mean	Bound	T	P
14	23000	1113	297	23527	-3.03	0.005

Notice that MINITAB gives the test statistic and the P-value, so that you can make your conclusion using either value. Since the P-value is smaller than α, you should Reject the null hypothesis.

> ▶ Exercise 29 (pg. 405) Soda consumed by teen-age males

Open worksheet **Ex7_3-29,** which is found in the **ch07** MINITAB folder. Click
on **Stat → Basic Statistics →1-Sample t.** Enter C1 for **Samples in columns,**
click on **Perform hypothesis test** and the **Hypothesized mean** is 3. Since the
claim is "teenage males drink less than 3 servings", you will perform a left-tailed
test. Click on **Options** and then on the down arrow beside **Alternative** to select
"less than". Click on **OK** twice and the results of the test should be displayed in
the Session Window.

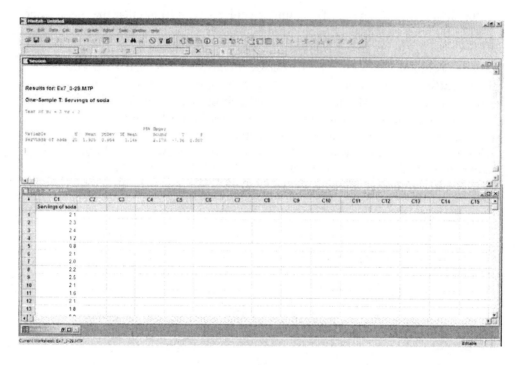

Since p=0.00, you would reject the null hypothesis. The data support the claim
that teenage males drink fewer than 3 12-oz. servings of soda per day.

► Exercise 30 (pg. 405) School supplies

Open worksheet **Ex7_3-30,** which is found in the **ch07** MINITAB folder.   Click
on **Stat → Basic Statistics →1-Sample t.** Enter C1 for **Samples in columns,**
click on **Perform hypothesis test** and the **Hypothesized mean** is 550. Since the
claim is "teachers spend more than $550 of their own money on school supplies
per year", you will perform a right-tailed test. Click on **Options** and then on the
down arrow beside **Alternative** to select "greater than". Click on **OK** and the
results of the test should be displayed in the Session Window.

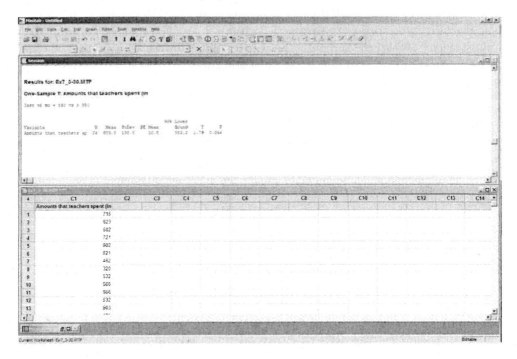

Since p=0.044, you should reject the null hypothesis. There is sufficient
evidence to show that teachers spend more than $550 on school supplies.

Section 7.4

> **Example 2 (pg. 409)** Hypothesis Test for a Proportion

Of 200 Americans, 49% are in favor of outlawing cigarettes. At $\alpha = .05$, is there enough evidence to reject the claim that 45% of Americans favor outlawing cigarettes?

Click on **Stat** $\rightarrow$ **Basic Statistics** $\rightarrow$ **1-Proportion.** The data is given in a summarized form, so select **Summarized data.** Enter 200 for the **Number of trials.** Since 49% of the sample were in favor, the **Number of events** is 98 (.49 * 200). Click on **Perform hypothesis test** and enter .45 for **Hypothesized proportion.**

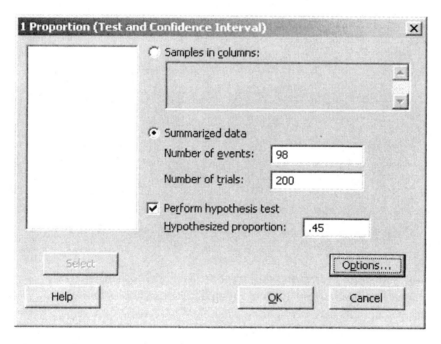

Click on **Options** and select "not equal" for the **Alternative.** Since np and nq are both larger than 5, click on **Use test and interval based on normal distribution,** and then click on **OK** twice.

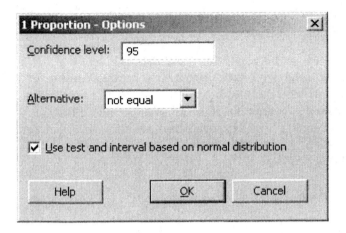

The results should be displayed in the Session Window.

Test and CI for One Proportion

Test of p = 0.45 vs p not = 0.45

Sample	X	N	Sample p	95% CI	Z-Value	P-Value
1	98	200	0.490000	(0.420719, 0.559281)	1.14	0.256

Using the normal approximation.

Notice that the test statistic (z = 1.14), the P-value (P = .256) and a 95% confidence interval for the true proportion of Americans in favor of outlawing cigarettes are all displayed in the output. With such a large P-value, you should fail to reject the null hypothesis.

> ▶ Exercise 11 (pg. 411) Environmentally conscious consumers

Click on **Stat → Basic Statistics → 1-Proportion.** The data is given in a summarized form, so select **Summarized data**. Enter 1050 for the **Number of trials.** Since 32% of the sample have stopped buying this product because of pollution concerns, the **Number of events** is 336 (.32 * 1050 = 336). Click on **Perform hypothesis test** and enter .30 for **Hypothesized proportion.** Click on **Options** and select "greater than" for the **Alternative.** Click on **Use test and interval based on normal distribution,** and then click on **OK** twice.

Test and CI for One Proportion

Test of p = 0.3 vs p > 0.3

Sample	X	N	Sample p	95% Lower Bound	Z-Value	P-Value
1	336	1050	0.320000	0.296321	1.41	0.079

Using the normal approximation.

Since p=0.079 and is greater than significance level 0.03, you would fail to reject the null hypothesis.

▸ Exercise 13 (pg. 412) Finding a Real Estate Agent

Click on **Stat → Basic Statistics → 1-Proportion.** The data is given in a
summarized form, so select **Summarized data**. Enter 1762 for the **Number of
trials.** Since 722 home buyers in the sample found their real estate agent through
a friend, the **Number of events** is 722. Click on **Perform hypothesis test** and
enter .44 for **Hypothesized proportion.** Click on **Options** and select "not equal"
for the **Alternative**. Click on **Use test and interval based on normal
distribution,** and then click on **OK** twice.

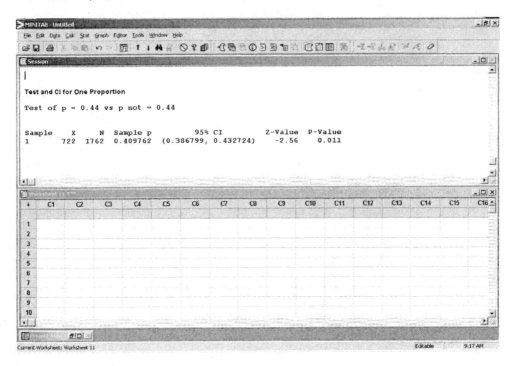

Since p=0.011 is less than significance level 0.02, you would reject the null
hypothesis.

Section 7.5

▶ Example 4 (pg. 417) Hypothesis test for a variance

Click on **Stat → Basic Statistics → 1-Variance.** First select **Enter Variance** in the top drop-down since this problem is given in terms of the variance. The data is given in a summarized form, so select **Summarized data**. Enter 41 for the **Sample size.** Enter .27 for the **Sample variance.** Click on **Perform hypothesis test** and enter .25 for **Hypothesized variance.** Click on **Options** and select "greater than" for the **Alternative** and then click on **OK** twice.

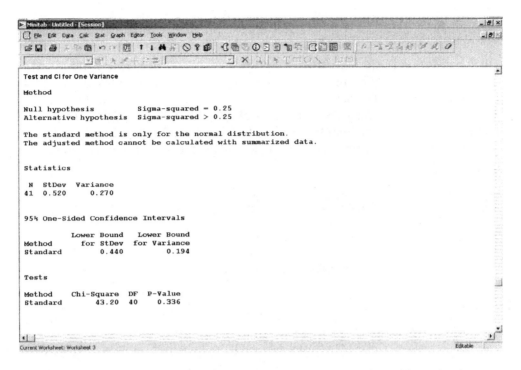

The Chi-square test statistic is 43.2 and the p-value is 0.336. With such a large p-value, you would fail to reject the null hypothesis.

> **Example 5 (pg. 418)**   Hypothesis test for a Standard deviation

Click on **Stat → Basic Statistics → 1-Variance.** First select **Enter Standard deviation** in the top drop-down since this problem is given in terms of the standard deviation. The data is given in a summarized form, so select **Summarized data**. Enter 23 for the **Sample size.** Enter .2.1 for the **Sample variance.** Click on **Perform hypothesis test** and enter 2.9 for **Hypothesized variance.** Click on **Options** and select "less than" for the **Alternative** and then click on **OK** twice.

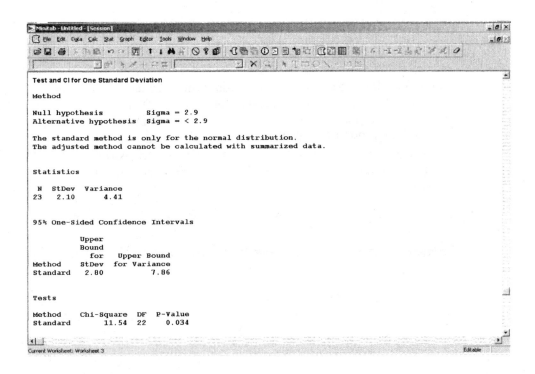

The Chi-square test statistic is 11.54 and the p-value is 0.034. With such a small p-value, you would reject the null hypothesis.

▶ Technology Lab (pg. 433) The Case of the Vanishing Women

4. Click on **Stat → Basic Statistics → 1-Proportion.** The data is given in a summarized form, so select **Summarized data**. Enter 100 for the **Number of trials.** Since 9 women were selected, the **Number of events** is 9. Click on **Options**. Enter .2914 for the **Test Proportion** because 29.14% of the original sample were women, and select "not equal" for the **Alternative**. Click on **Use test and interval based on normal distribution,** and then click on **OK** twice.

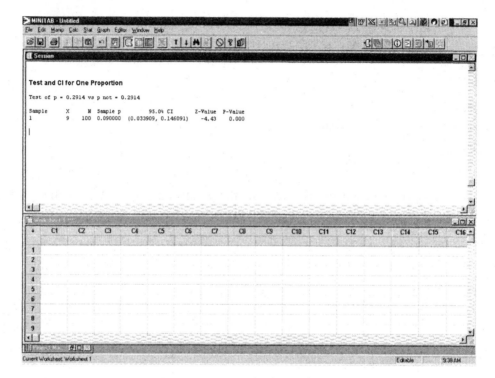

Hypothesis Testing with Two Samples

CHAPTER

8

Section 8.1

▶ Exercise 29 (pg. 447)   Time spent watching TV

Open worksheet **Ex8_1-29.mtp**, which is found in the **ch08** MINITAB folder. The 1981 data (Time A) is in C1 and the new data (Time B) is in C2. Notice that for both samples, n = 30. MINITAB does not have a 2-sample Z-test, but you can use a 2-sample t-test instead since the t distribution becomes very similar to the normal distribution as the sample size approaches 30. Click on **Stat → Basic Statistics → 2-Sample t.** Select **Samples in different columns** and enter C1 for the **First** and C2 for the **Second** column.

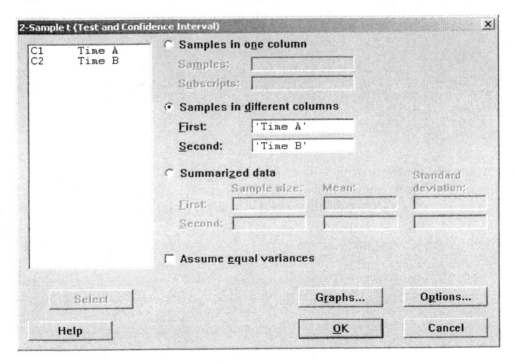

Click on **Options**, and then on the down arrow beside **Alternative** and select **greater than** since the sociologist claims that children spent more time watching TV in 1981 than they do today. Be sure that **Test difference** is 0.

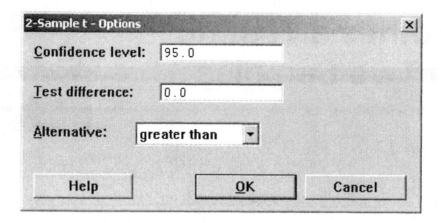

Click on **OK** twice and the results of the test should be displayed in the Session Window.

Two-Sample T-Test and CI: Time A, Time B

```
Two-sample T for Time A vs Time B

          N    Mean   StDev   SE Mean
Time A   30   2.130   0.490    0.089
Time B   30   1.757   0.470    0.086


Difference = mu (Time A) - mu (Time B)
Estimate for difference:   0.373
95% lower bound for difference:   0.166
T-Test of difference = 0 (vs >): T-Value = 3.01   P-Value = 0.002
DF = 57
```

Notice that the test statistic is T = 3.01 with a P-value = 0.002. Since this P-value is so small, you would Reject H$_0$ at any α level. Thus, the sociologist's claim is true – children watched more TV in 1981.

◀

▶ Exercise 31 (pg. 448)   Difference between Washer Diameters

Open worksheet **Ex8_1-31.mtp**, which is found in the **ch08** MINITAB folder. The diameters from the first method are in C1 and the diameters from the second method are in C2. Notice that for both samples, n = 35. MINITAB does not have a 2-sample Z-test, but you can use a 2-sample t-test instead since the t distribution becomes very similar to the normal distribution as the sample size gets larger than 30. Click on **Stat → Basic Statistics → 2-Sample t.** Select **Samples in different columns** and enter C1 for the **First** and C2 for the **Second** column. Click on **Options**, and then on the down arrow beside **Alternative** and select **not equal** since the production engineer claims there is no difference between the two methods. Click on **OK** twice and the results of the test should be displayed in the Session Window.

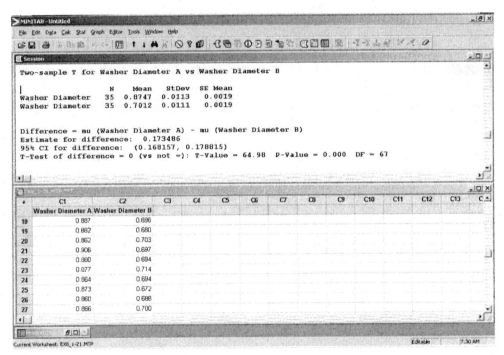

Since the P-value is so small, you would Reject H_0 at any α level.

> ▸ Exercise 32 (pg. 448)   Difference between Nut Diameters

Open worksheet **Ex8_1-32.mtp**, which is found in the **ch08** MINITAB folder.
The diameters from the first method are in C1 and the diameters from the second
method are in C2. Notice that for both samples, n = 40. MINITAB does not
have a 2-sample Z-test, but you can use a 2-sample t-test instead since the t
distribution becomes very similar to the normal distribution as the sample size
gets larger than 30. Click on **Stat → Basic Statistics → 2-Sample t.** Select
Samples in different columns and enter C1 for the **First** and C2 for the **Second**
column. Click on **Options**, and then on the down arrow beside **Alternative** and
select **not equal** since the production engineer claims there is no difference
between the two methods. Click on **OK** twice and the results of the test should be
displayed in the Session Window.

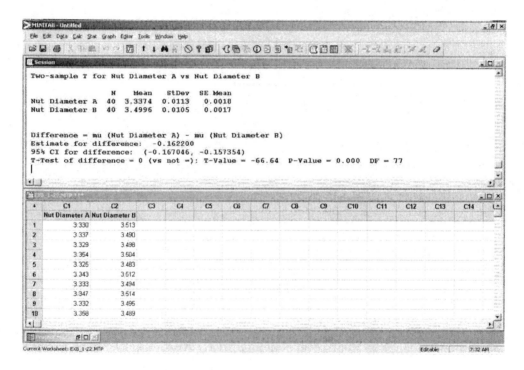

Since the P-value is so small, you would Reject H_0 at any α level.

Section 8.2

> **Example 1 (pg. 454)** Braking Distance

Braking distance was tested for 8 Volkwagon GTIs and 10 Ford Focuses when traveling on dry pavement at 60 mph. The summarized data is shown on the left of page 454 in the text. Click on **Stat → Basic Statistics → 2-Sample t.** Select **Summarized data** and enter the Volkswagon data for the **First** sample and the Ford data for the **Second** sample. Click on **Options,** and then on the down arrow beside **Alternative** and select **not equal** since you want to test whether the mean stopping distances are different.

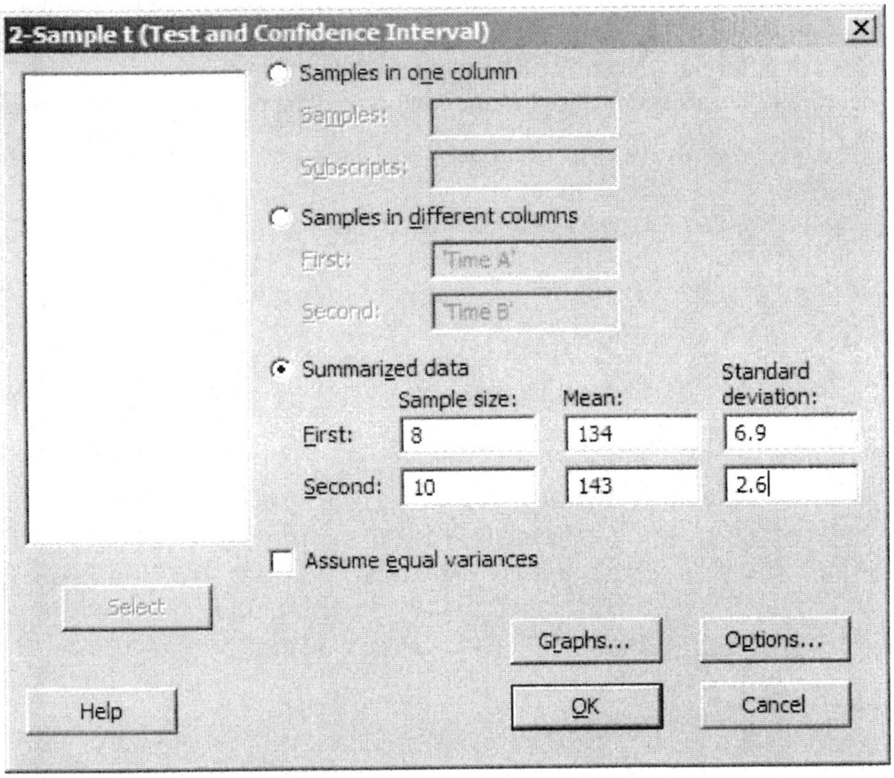

Click on **OK** twice and the results of the test should be displayed in the Session Window.

Two-Sample T-Test and CI

```
Sample   N    Mean   StDev  SE Mean
1        8  134.00   6.90      2.4
2       10  143.00   2.60     0.82


Difference = mu (1) - mu (2)
Estimate for difference:  -9.00
95% CI for difference:  (-14.94, -3.06)
T-Test of difference = 0 (vs not =): T-Value = -3.50   P-Value =
0.008  DF = 8
```

Notice that the test statistic is T= -3.5. Since the P-value = 0.008 and is less than the α-level of .01, there is evidence to conclude that the mean braking distances of the tires are different. (Note that Minitab uses 8 degrees of freedom, rather than 7 degrees of freedom as is used in the textbook. The p-value will be slightly different.)

> **Exercise 21 (pg. 458)** Tensile Strength of Steel Bars

Open worksheet **Ex8_2-21.mtp**, which is found in the **ch08** MINITAB folder.
The New method data is in C1 and the Old method data is in C2. Click on **Stat
→ Basic Statistics → 2-Sample t.** Select **Samples in different columns** and
enter C1 for the **First** and C2 for the **Second** column. Select **Assume Equal
Variances,** since the problem tells you to assume the population variances are
equal. Click on **Options**, and then on the down arrow beside **Alternative** and
select **not equal** since you want to test if the new treatment makes a difference in
the tensile strength of steel bars. Click on **OK** twice and the results of the test
should be displayed in the Session Window.

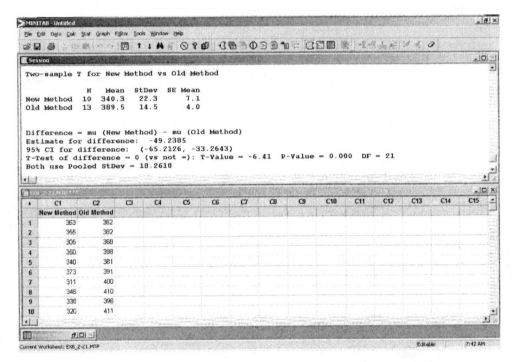

> ## ▶ Exercise 24 (pg. 459) Comparing Teaching Methods

Open worksheet **Ex8_2-24.mtp**, which is found in the **ch08** MINITAB folder.
The Traditional Lab data is in C1 and the Interactive data is in C2. Click on **Stat**
→ **Basic Statistics** → **2-Sample t.** Select **Samples in different columns** and
enter C1 for the **First** and C2 for the **Second** column. Select **Assume Equal
Variances,** since the problem tells you to assume the population variances are
equal. Click on **Options**, and then on the down arrow beside **Alternative** and
select **less than** since you want to test if the students taught in a traditional lab
had lower science test scores than students taught with the interactive simulation
software. Click on **OK** twice and the results of the test should be displayed in the
Session Window.

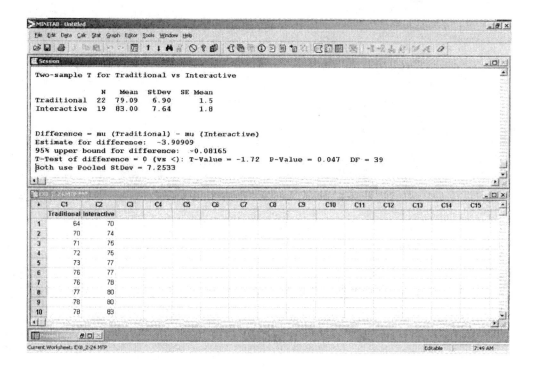

Section 8.3

▶ **Example 1 (pg. 463)** Golf Scores

Enter the data, found on page 463 of the textbook, into the MINITAB Data Window. Put the Old Design data in C1 and the New Design data in C2. Click on **Stat → Basic Statistics → Paired t.** Enter C1 for the **First Sample** and C2 for the **Second Sample.**

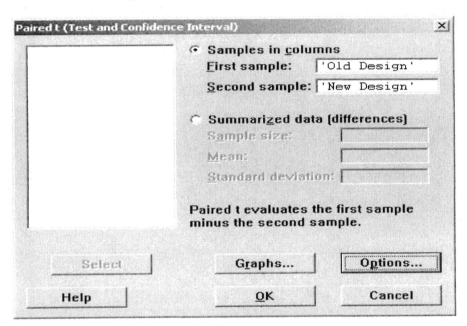

Click on **Options.** Enter 0 for **Test Mean** and select **greater than** as the **Alternative.** Click on **OK** twice to display the results.

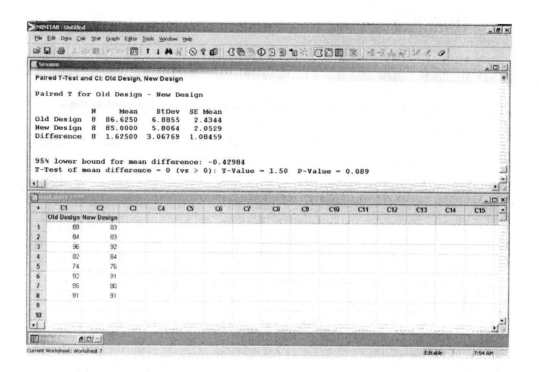

▶ Exercise 9 (pg. 466) Verbal SAT Scores

Open worksheet **Ex8_3-9.mtp**, which is found in the **ch08** MINITAB folder. The scores on the first SAT are in C2 and the scores on the second SAT are in C3. Click on **Stat → Basic Statistics → Paired t.** Enter C1 for the **First Sample** and C2 for the **Second Sample.** Click on **Options.** Enter 0 for **Test Mean** and select **less than** as the **Alternative** because, if the scores have improved, then the difference (first SAT - second SAT) will be less than 0. Click on **OK** twice to display the results.

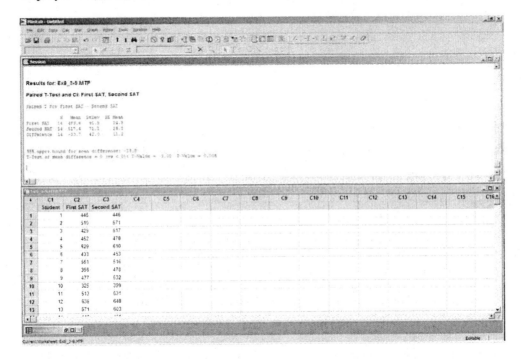

Since p=.005, you would reject the null hypothesis.

▶ Exercise 17 (pg. 469) Does new drug reduce blood pressure?

Enter the "before" blood pressures are in C1 and the "after" blood pressures in
C2. Click on **Stat** → **Basic Statistics** → **Paired t.** Enter C1 for the **First
Sample** and C2 for the **Second Sample.** Click on **Options.** Enter 0 for **Test
Mean** and select **greater than** as the **Alternative** because if the new drug
reduces blood pressure, then the difference (before - after) will be greater than 0.
Click on **OK** twice to display the results.

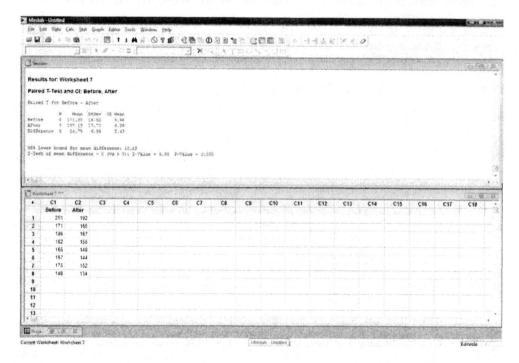

Since the P-value is so small, you should Reject the null hypothesis. Thus, the
new drug appears to reduce blood pressure.

Section 8.4

> ▶ **Example 1 (pg. 473)** Difference between male and female
> Internet Users

In a study of 200 female and 250 male Internet users, 30% of the females and
38% of the males plan to shop on-line. This is a summary of the results of the
study. To test if there is a difference in the proportion of male and female users
who plan to shop on-line, click on **Stat → Basic Statistics → 2 Proportions.**
Select **Summarized Data** and use the data for Females as the **First sample**.
Enter 200 **Trials** and 60 **Events** (200 x .30). Use the data for Males as the
Second sample. Enter 250 **Trials** and 95 **Events** (250 x .38).

2 Proportions (Test and Confidence Interval)		✕
	○ **Samples in one column**	
	Samples: [ ]	
	Subscripts: [ ]	
	○ **Samples in different columns**	
	First: [ ]	
	Second: [ ]	
	● **Summarized data**	
		Trials:	Events:
	First:	200	60
	Second:	250	95
Select		Options...
Help	OK	Cancel

Click on **Options.** Enter 0 for **Test mean**, and select **not equal** as the
Alternative since you want to test if there is a difference between the proportion
of male and female shoppers. Next click on **Use pooled estimate of p for test.**

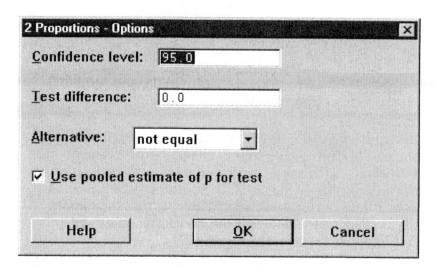

Click on **OK** twice to display the results in the Session Window.

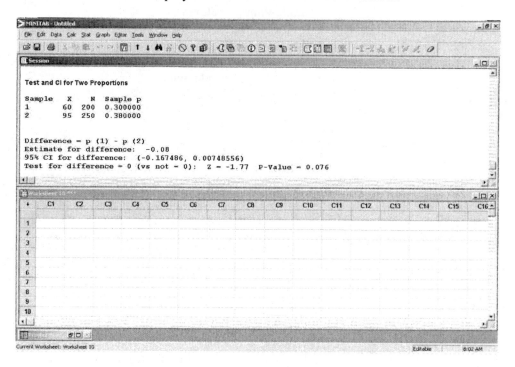

Since the P-value is smaller than α, you should Reject the null hypothesis.

▶ Exercise 7 (pg. 475) Alternative Medicine Usage

To test if there is a difference in the proportion of adults who used alternative medicine in 1991 and the proportion of adults who use it now, click on **Stat → Basic Statistics → 2 Proportions.** Select **Summarized Data** and use the data for 1991 as the **First sample.** Enter 1539 **Trials** and 520 **Events.** Use the recent study data as the **Second sample.** Enter 2055 **Trials** and 865 **Events.** Click on **Options.** Enter 0 for **Test mean**, and select **not equal** as the **Alternative** since you want to test if there is a difference between the proportion of users in 1991 and the present. Next click on **Use pooled estimate of p for test.**
Click on **OK** twice to display the results in the Session Window.

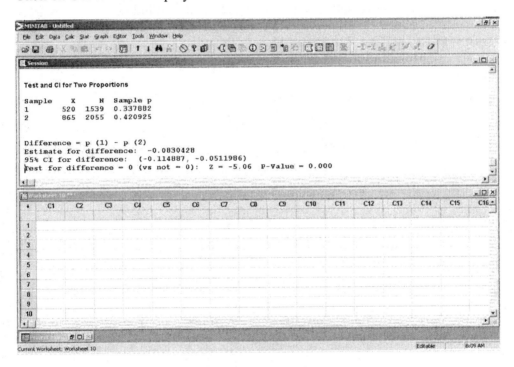

With such a small P-value, you should Reject the null hypothesis. Thus, the proportion of adults using alternative medicines has changed since 1991.

> ► Exercise 12 (pg. 476) Fewer Smokers in California?

To test if the proportion of adult smokers in California is lower than the proportion of adult smokers in Oregon, click on **Stat → Basic Statistics → 2 Proportions.** Select **Summarized Data** and use the data for California as the **First sample**. Enter 228 **Events** (1500 x .152) and 1500 **Trials**. Use the Oregon data as the **Second sample**. Enter 185 **Events** (1000 x .185) and 1000 **Trials.** Click on **Options**. Enter 0 for **Test mean**, and select **less than** as the **Alternative** since you want to test if the proportion of smokers in California is lower than the proportion in Oregon. Next click on **Use pooled estimate of p for test.**
Click on **OK** twice to display the results in the Session Window.

Test and CI for Two Proportions

```
Sample    X     N    Sample p
1        228   1500   0.152000
2        185   1000   0.185000


Difference = p (1) - p (2)
Estimate for difference:  -0.033
95% upper bound for difference:  -0.00769353
Test for difference = 0 (vs < 0):  Z = -2.18   P-Value = 0.015

Fisher's exact test: P-Value = 0.017
```

Since p=0.017, you would reject the null hypothesis and say that there is enough evidence that California has fewer smokers.

> **Technology Lab (pg. 487)** Tails Over Heads

1. Click on **Stat → Basic Statistics → 1 Proportion.** Select **Summarized Data** and enter 11902 **Trials** and 5772 **Events.** Click on **Options.** Enter .5 for **Test proportion**, and select **not equal** as the **Alternative** since you want to test if the probability is .5 or not. Next click on **Use test and interval based on normal distribution.** Click on **OK** twice to display the results in the Session Window.

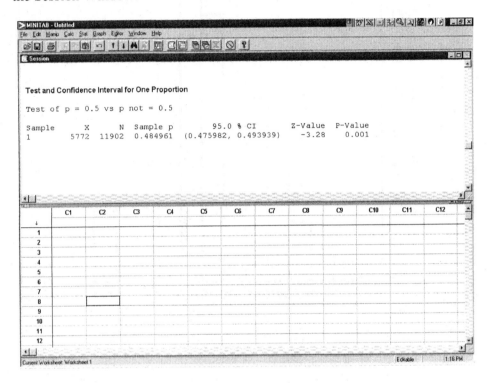

3. To repeat the simulation, click on **Calc → Random data → Binomial.** You want to **Generate** 500 **rows of data** and **Store In Column** C1. The **Number of trials** is 11902 and the **Probability of success** is .5. Click on **OK** and C1 should have 500 rows of data in it. To draw the histogram, click on **Graph → Histogram → Simple.** Enter C1 for the **Graph variable.** Click on **OK** to view the histogram.

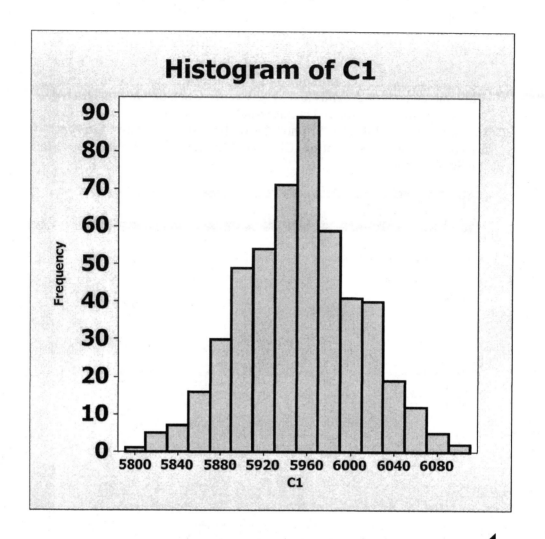

Correlation and Regression

CHAPTER

9

Section 9.1

▶ Example 3 (pg. 498)   Constructing a Scatter Plot

Open worksheet **OldFaithful,** which is found in the **ch09** MINITAB folder. The
duration (in minutes) of several of Old Faithful's eruptions should be in C1, and
the time (in minutes) until the next eruption should be in C2. Notice that
Duration is the x-variable and Time is the y-variable. To plot the data, click on
Graph → Scatterplot → Simple. On the input screen, enter C2 for the **Y
variable** and C1 for the **X variable.**

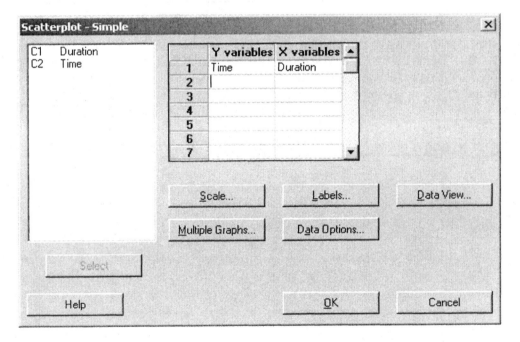

Next, click on **Labels** and enter a title for the plot. Click on OK to view the plot
produced using Minitab default settings.

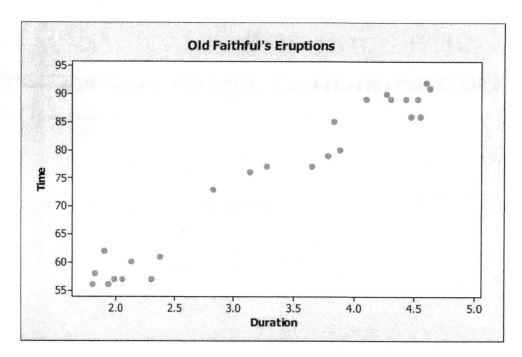

Click on **OK** twice to view the scatter plot. This plot would be better if both axes started at 0. To make this change, right-click on the X-axis scale and select "Edit X-scale". Enter 0 : 5 / .5 for **Position of ticks**. Click on **OK**. Next right-click on the Y-axis scale and select "Edit Y-scale". Enter 0 : 100 / 10 for **Position of ticks**. Click on **OK** to view the completed plot.

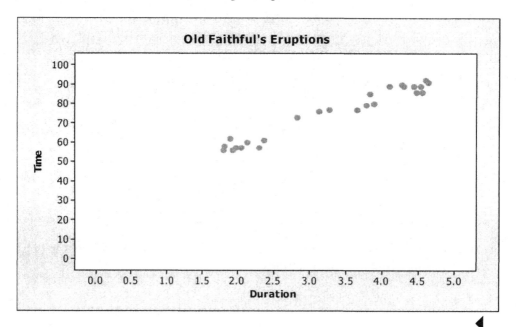

► Example 5 (pg. 501) Finding the Correlation Coefficient

Open worksheet **OldFaithful,** which is found in the **ch09** MINITAB folder. The
duration (in minutes) of several of Old Faithful's eruptions should be in C1, and
the time (in minutes) until the next eruption should be in C2. Notice that
Duration is the x-variable and Time is the y-variable. To find the correlation
coefficient, click on **Stat → Basics Statistics → Correlation.** On the input
screen, select both C1 and C2 for **Variables,** by double-clicking on each one.

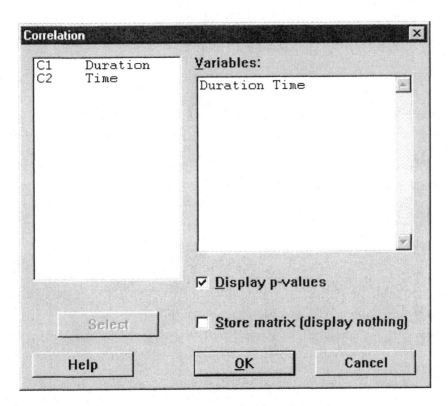

Click on **OK** and the Correlation Coefficient will be displayed in the Session
Window.

Correlations: Duration, Time

```
Pearson correlation of Duration and Time = 0.979
P-Value = 0.000
```

▸ Exercise 17 (pg. 508) Plot of Study Hours vs. Test Scores

Open worksheet **Ex9_1-17,** which is found in the **ch09** MINITAB folder. The hours spent studying should be in C1, and the test scores should be in C2. Notice that "Hours study" is the x-variable and "Test Scores" is the y-variable. To plot the data, click on **Graph → Scatterplot → Simple.** On the input screen, enter C2 for the **Y variable** and C1 for the **X variable.** Next, click on **Labels** and enter a title for the plot. Click on OK to view the plot produced using Minitab default settings. Edit the plot to set the tick mark positions so that both axes begin at zero. One possibility is to enter the tick positions for the **X variable** as 0:8/1 (0 to 8 in steps of 1) and for the **Y variable** as 0:100/10 (0 to 100 in steps of 10). Click on **OK** twice to view the scatter plot.

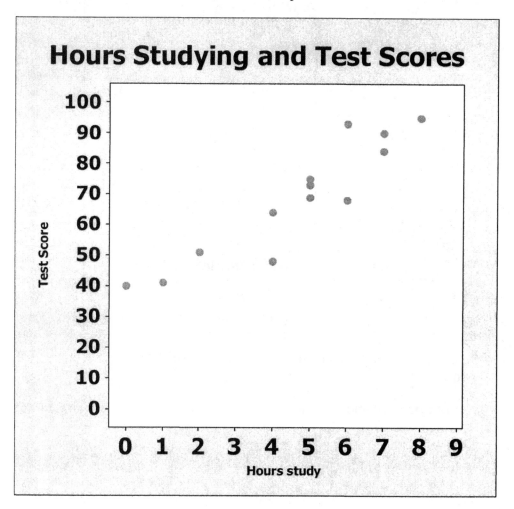

Now, to find the correlation coefficient, click on **Stat → Basics Statistics → Correlation.** On the input screen, select both C1 and C2 for **Variables,** by

double-clicking on each one. Click on **OK** and the output should be in the
Session Window.

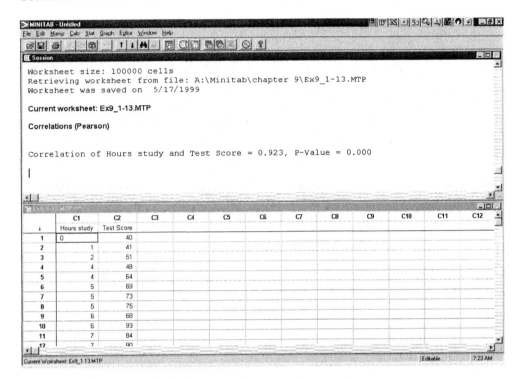

Section 9.2

Example 2 (pg. 515) Finding a Regression Equation

Open worksheet **OldFaithful,** which is found in the **ch09** MINITAB folder. The duration (in minutes) of several of Old Faithful's eruptions should be in C1, and the time (in minutes) until the next eruption should be in C2. Notice that Duration is the x-variable and Time is the y-variable. To find the regression equation, click on **Stat → Regression → Regression.** Enter C2 (Time) for the **Response** variable, and C1 (Duration) as the **Predictor.**

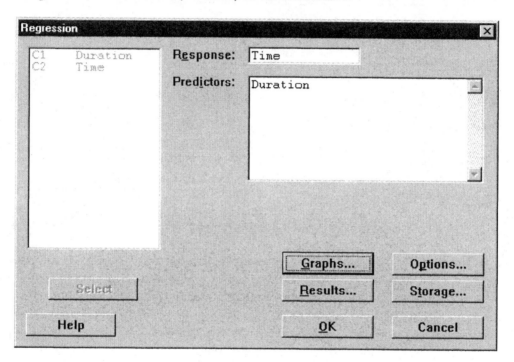

Click on **Results.** Select **Regression equation, table of coefficients, s, R-squared, and basic analysis of variance.**

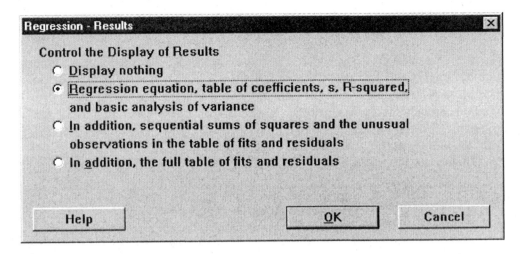

Click on **OK** twice to view the output in the Session Window.

Regression Analysis: Time versus Duration

```
The regression equation is
Time = 33.7 + 12.5 Duration


Predictor       Coef    SE Coef      T       P
Constant      33.683      1.894   17.79   0.000
Duration     12.4809      0.5464  22.84   0.000


S = 2.88153    R-Sq = 95.8%    R-Sq(adj) = 95.6%


Analysis of Variance

Source           DF       SS       MS       F       P
Regression        1   4331.7   4331.7  521.69   0.000
Residual Error   23    191.0      8.3
Total            24   4522.6
```

Notice that the regression equation is Time = 33.7 + 12.5 * Duration.

> ▸ Exercise 16 (pg. 518)   Hours Online vs. Test scores

Open worksheet **Ex9_2-16,** which is found in the **ch09** MINITAB folder. Hours
Online is in C1 and Test is in C2. First plot the data. To plot the data, click on
Graph → Scatterplot → Simple. On the input screen, enter C2 for the **Y
variable** and C1 for the **X variable.** Next, click on **Labels** and enter a title for
the plot. Click on OK to view the plot produced using Minitab default settings.
Edit the plot to set the tick mark positions so that both axes begin at zero. Click
on **OK** to view the scatter plot.

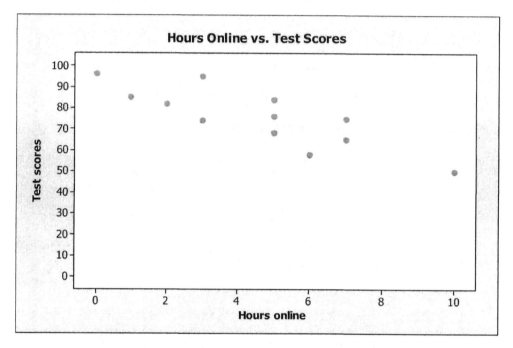

To find the regression equation, click on **Stat → Regression → Regression.**
Enter C2 for the **Response** variable, and C1 as the **Predictor.** Click on **Results.**
Select **Regression equation, table of coefficients, s, R-squared, and basic
analysis of variance.** Click on **OK** twice.

Regression Analysis: Test scores versus Hours online

The regression equation is
Test scores = 94.0 - 4.07 Hours online

Predictor	Coef	SE Coef	T	P
Constant	93.970	4.524	20.77	0.000
Hours online	-4.0674	0.8600	-4.73	0.001

S = 8.11334   R-Sq = 69.1%   R-Sq(adj) = 66.0%

Analysis of Variance

Source	DF	SS	MS	F	P
Regression	1	1472.4	1472.4	22.37	0.001
Residual Error	10	658.3	65.8		
Total	11	2130.7			

Section 9.3

▶ Example 2 (pg. 528)   Finding the Standard Error and the
Coefficient of Determination

Enter the first two columns of data (found on page 528 of the textbook) into the
MINITAB Data Window. Enter the x's into C1 and name it Expenses. Enter the
y's into C2 and name it Sales. Both the coefficient of determination and the
standard error of the estimate are part of the regression output. To find the
regression equation, click on **Stat → Regression → Regression.** Enter C2 for
the **Response** variable, and C1 as the **Predictor.** Click on **Results.** Select
**Regression equation, table of coefficients, s, R-squared, and basic analysis of
variance.** Click on **OK** twice.

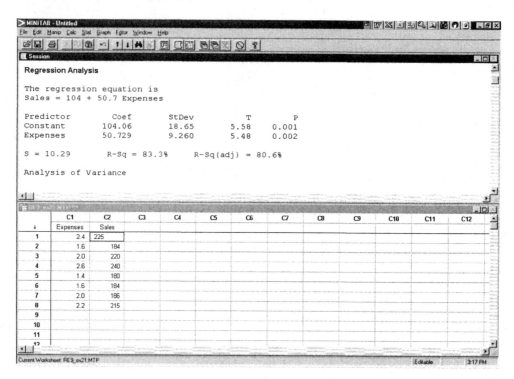

Notice that the standard error of the estimate is S = 10.29 and the coefficient of
determination is R-Sq = 83.3%.

▶ Example 3 (pg. 530)   Constructing a Prediction Interval

Enter the first two columns of data (found on page 528 of the textbook) into the MINITAB Data Window. Enter the x's into C1 and name it **Expenses.** Enter the y's into C2 and name it **Sales.** To find the regression equation, click on **Stat →
Regression → Regression.** Enter C2 for the **Response** variable, and C1 as the **Predictor.** Now, to find both the point estimate and the prediction interval, click on **Options. (Fit Intercept** is selected by default.) Next enter the advertising expenditure. Although the amount given in this problem is $2100, the data is stored in thousands of dollars. So enter 2.1 for **Prediction interval for new observations.** Enter 95 for the **Confidence level** and select **Prediction limits.**

Click on **OK**. Click on **Results.** If you would like to see the other regression output, then select **Regression equation, table of coefficients, s, R-squared, and basic analysis of variance.** If you only want to see the prediction interval, then select **Display nothing.** Click on **OK** twice and view the output in the Session Window.

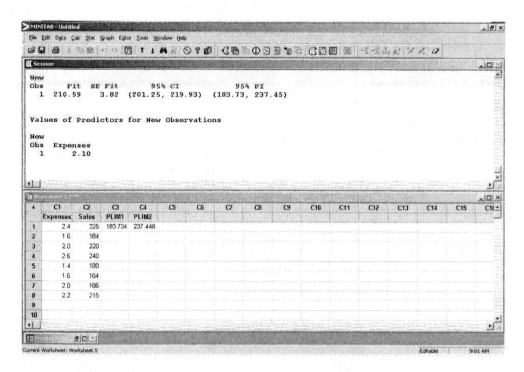

Notice that the predicted value, 210.59, is listed below **Fit** and the prediction interval, (183.73, 237.45), is listed below **95% PI.**

> Exercise 11 (pg. 532) and Exercise 19 (pg. 533) Retail space
>   vs. Sales

Open worksheet **Ex9_3-11,** which is found in the **ch09** MINITAB folder. Square
footage is in C1 and Sales is in C2. For these two problems, you need to find the
coefficient of determination, the standard error of the estimate, and a 90%
prediction interval when square footage is 4.5 billion. This can be accomplished
all at once in MINITAB. Click on **Stat** → **Regression** → **Regression.** Enter C2
for the **Response** variable, and C1 as the **Predictor.** Now, to find both the point
estimate and the prediction interval, click on **Options.** Select **Fit Intercept** by
clicking on it. Next enter the square footage. So enter 4.5 for **Prediction
interval for new observations.** Enter 90 for the **Confidence level** and select
Prediction limits. Click on **OK**. Click on **Results.** Since you would like to see
the other regression output, then select **Regression equation, table of
coefficients, s, R-squared, and basic analysis of variance.** Click on **OK** twice
and view the output in the Session Window.

Regression Analysis: Sales versus Square footage

```
The regression equation is
Sales = - 1882 + 549 Square footage

Predictor          Coef   SE Coef       T      P
Constant        -1881.7     143.8  -13.09  0.000
Square footage   549.45     25.79   21.30  0.000

S = 30.5759    R-Sq = 98.1%    R-Sq(adj) = 97.8%

Analysis of Variance

Source          DF      SS      MS       F      P
Regression       1  424297  424297  453.85  0.000
Residual Error   9    8414     935
Total           10  432711

Predicted Values for New Observations

New
Obs     Fit  SE Fit      90% CI              90% PI
  1  590.82   28.94  (537.77, 643.87)  (513.65, 668.00)X

X denotes a point that is an outlier in the predictors.

Values of Predictors for New Observations

New    Square
Obs    footage
  1       4.50
```

Section 9.4

> ▶ Example 1 (pg. 536) Finding a Multiple Regression Equation

Open worksheet **Salary,** which is found in the **ch09** MINITAB folder.
Salary should be in C1, Employment in C2, Experience in C3, and Education in
C4. Click on **Stat** → **Regression** → **Regression.** Enter C1 for the **Response**
variable, and enter C2, C3, and C4 as the **Predictors.** Click on **Results.** Since
you would like to see the other regression output, then select **Regression
equation, table of coefficients, s, R-squared, and basic analysis of variance.**
Click on **OK** twice and view the output in the Session Window.

Regression Analysis: Salary versus Employment (, Experience (, ...

```
The regression equation is
Salary = 49764 + 364 Employment (yrs) + 228 Experience (yrs)
        + 267 Education (yrs)


Predictor           Coef   SE Coef      T      P
Constant           49764      1981  25.12  0.000
Employment (yrs)   364.41     48.32   7.54  0.002
Experience (yrs)   227.6      123.8   1.84  0.140
Education (yrs)    266.9      147.4   1.81  0.144


S = 659.490   R-Sq = 94.4%   R-Sq(adj) = 90.2%


Analysis of Variance

Source          DF        SS       MS      F      P
Regression       3  29231989  9743996  22.40  0.006
Residual Error   4   1739708   434927
Total            7  30971697
```

The regression equation is listed at the beginning of the output. Notice that the
regression equation uses the values listed below **Coef.** These values are the
coefficients of the multiple regression equation.

◀

▶ Exercise 5 (pg. 540) Finding a Multiple Regression Equation

Open worksheet **Ex9_4-5,** which is found in the **ch09** MINITAB folder.
Sales should be in C1, Square footage in C2, and Number of shopping centers in
C3. Click on **Stat → Regression → Regression.** Enter C1 for the **Response**
variable, and enter C2 and C3 as the **Predictors.** Click on **Results.** Since you
would like to see the other regression output, then select **Regression equation,
table of coefficients, s, R-squared, and basic analysis of variance.** Click on
OK twice and view the output in the Session Window.

Regression Analysis: Sales versus Square footage, Shopping centers

```
The regression equation is
Sales = - 2518 + 127 Square footage + 66.4 Shopping centers


Predictor              Coef   SE Coef      T      P
Constant            -2518.4     435.0  -5.79  0.000
Square footage        126.8     275.8   0.46  0.658
Shopping centers      66.36     43.13   1.54  0.162


S = 28.4890   R-Sq = 98.5%   R-Sq(adj) = 98.1%


Analysis of Variance

Source          DF       SS       MS       F      P
Regression       2   426218   213109  262.57  0.000
Residual Error   8     6493      812
Total           10   432711
```

▶ Technology Lab (pg. 549) Sugar, Fat, and Carbohydrates

Open worksheet **Tech9,** which is found in the **ch09** MINITAB folder.
Cereal Name is in C1, Calories is in C3, Sugar is in C4, Fat is in C5, and
Carbohydrates is in C6.

1. Click on **Graph → Scatterplot → Simple.** On the input screen, enter C3 for
    the **X variable** and C4 for the **Y variable.** Click on **Labels** and enter a title.
    Click on **OK.** Repeat these steps for parts b - f, changing the variables as
    directed in each part.

3. Click on **Stat → Basics Statistics → Correlation.** On the input screen,
    select both C3 and C4 for **Variables** by double-clicking on each one. Click
    on **OK** and the output should be in the Session Window. Repeat these steps
    for each pair of variables listed in question 1, parts b - f.

4. (Do both 4 & 5 at one time here). Click on **Stat → Regression →
    Regression.** Enter C4 for the **Response** variable, and enter C3 as the
    Predictor. Next, to predict the sugar content of 1 cup of cereal with a calorie
    content of 120 kcal, click on **Options.** Select **Fit Intercept** by clicking on it.
    Next enter the calorie content. So enter 120 for **Prediction interval for new
    observations.** Enter 95 for the **Confidence level** and select **Prediction
    limits.** Click on **OK.** Next, click on **Results.** Since you would like to see
    the other regression output, select **Regression equation, table of
    coefficients, s, R-squared, and basic analysis of variance.** Click on **OK**
    twice and view the output in the Session Window. Repeat these steps using
    C5 for the **Response** variable.

5. (Do both 6 & 7 at one time here). Click on **Stat → Regression →
    Regression.** Enter C3 for the **Response** variable, and enter C4, C5, and C6
    as the **Predictors.** Next, click on **Results.** Since you would like to see the
    other regression output, then select **Regression equation, table of
    coefficients, s, R-squared, and basic analysis of variance.** Click on **OK**
    twice and view the output in the Session Window. For part b, repeat these
    steps using C4 and C6 for the **Predictors.** Next, to predict the calorie content
    of 1 cup of cereal with 10g of sugar and 25g of carbohydrates, click on
    Options. Select **Fit Intercept** by clicking on it. Next enter the sugar and
    carbohydrate contents. To enter both of these, type in a 10 (for the sugar),
    leave a space, and then type in a 25 (for the carbohydrate) for **Prediction
    interval for new observations.** Click on **OK.** Next, click on **Results.**
    Since you would like to see the other regression output, then select
    **Regression equation, table of coefficients, s, R-squared, and basic
    analysis of variance.** Click on **OK** twice and view the output in the Session
    Window.

Chi-Square Tests and the F-Distribution

CHAPTER

10

Section 10.1

▶ Example 3 (pg. 556)   The Chi-Square Goodness-of-Fit Test

Enter the data into the MINITAB Data Window. Enter the Responses into C1 and name it Response. Enter the observed frequencies into C2 and name it Observed. Enter the distribution (from the null hypothesis) into C3 and name it Distribution. These values should be proportions, not percentages. Thus, 44% should be entered as .44.

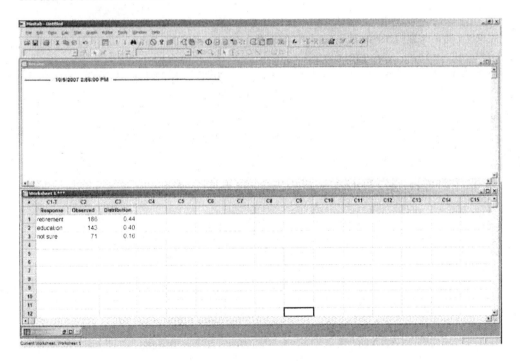

To perform the Goodness of Fit test, click on **Stat → Tables → Chi-square Goodness of Fit Test.**

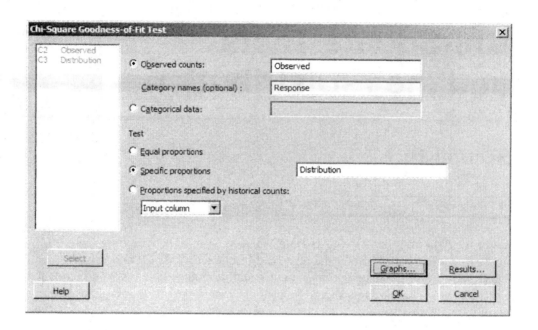

Select **Observed counts** and enter C2 (Observed). Click on the input box beside **Category names** and enter C1 (Response). Select **Specific proportions** and enter C3 (Distribution) into the input box. Click on **OK** and the Chi-Square test statistic will be displayed in the Session Window.

Chi-Square Goodness-of-Fit Test for Observed Counts in Variable: Observed

Using category names in Response

Category	Observed	Test Proportion	Expected	Contribution to Chi-Sq
retirement	186	0.44	176	0.56818
education	143	0.40	160	1.80625
not sure	71	0.16	64	0.76563

N	DF	Chi-Sq	P-Value
400	2	3.14006	0.208

In this example, the test statistic is 3.14006. The P-value is 0.208, which is larger than $\alpha = .05$. You should not reject the null hypothesis.

> Exercise 12 (pg. 562)   Bicycle Accidents

Open worksheet **10_1_12** found in the MINITAB folder **ch10.** The months
should be in C1, and the frequencies in C2. Can you conclude that accidents are
uniformly distributed by month?

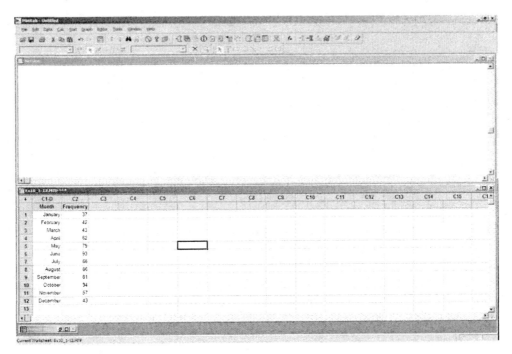

To perform the Goodness of Fit test, click on **Stat → Tables → Chi-square
Goodness of Fit Test.** Select **Observed counts** and enter C2 (Frequency).
Click on the input box beside **Category names** and enter C1 (Month). Select
Equal proportions since we are assuming a uniform distribution. Click on **OK**
and the Chi-Square test statistic will be displayed in the Session Window.

Chi-Square Goodness-of-Fit Test for Observed Counts in Variable: Frequency

Using category names in Month

Category	Observed	Test Proportion	Expected	Contribution to Chi-Sq
January	37	0.0833333	65.25	12.2308
February	42	0.0833333	65.25	8.2845
March	43	0.0833333	65.25	7.5872
April	62	0.0833333	65.25	0.1619
May	79	0.0833333	65.25	2.8975
June	93	0.0833333	65.25	11.8017
July	66	0.0833333	65.25	0.0086
August	86	0.0833333	65.25	6.5987
September	81	0.0833333	65.25	3.8017
October	94	0.0833333	65.25	12.6676
November	57	0.0833333	65.25	1.0431
December	43	0.0833333	65.25	7.5872

N	DF	Chi-Sq	P-Value
783	11	74.6705	0.000

In this example, the test statistic is 74.6705 and the P-value=0.000. Since this P-value is so small, you would reject the null hypothesis at any α-level.

◀

Section 10.2

▶ Example 3 (pg. 570) Chi-Square Independence Test

Enter the data into the MINITAB Data Window. First label the columns: use Gender for C1, "0-1" for C2, ... "6-7" for C5. Now enter the data into the appropriate columns. Do not enter any totals.

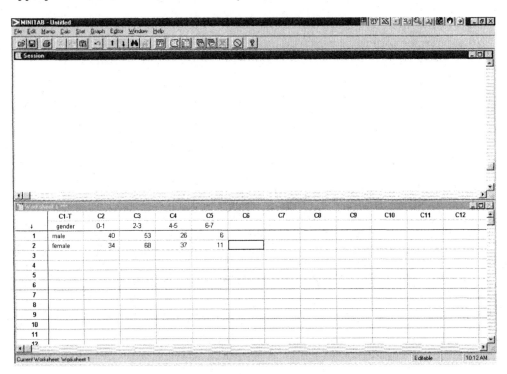

To perform the chi-square independence test, click on **Stat → Tables → Chi-square Test (Two way table in worksheet).** On the input screen, select C2 - C5 for the **Columns containing the table.** Click on **OK** and the test results will be displayed in the Session Window.

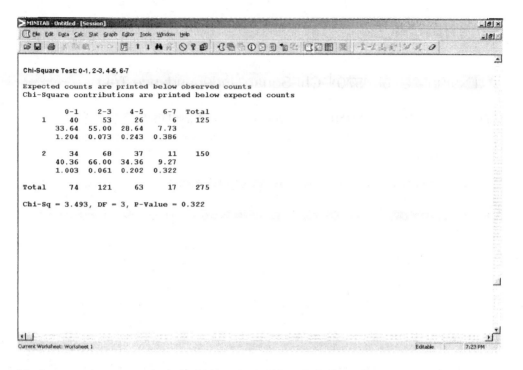

Notice that the test statistic is Chi-Sq = 3.493 and the P-value = .322. Since this P-value is larger than $\alpha = .05$, you should not reject the null hypothesis. Thus, there is not enough evidence to conclude that the number of days per week spent exercising is related to gender.

▶ Exercise 17 (pg. 573) Should the drug be used as treatment?

Enter the data into the MINITAB Data Window. Enter Result data into C1, Drug data into C2 and Placebo data into C3. To perform the chi-square independence test, click on **Stat → Tables → Chi-square Test (Two way table in worksheet).** On the input screen, select C2 - C3 for the **Columns containing the table.** Click on **OK** and the test results will be displayed in the Session Window.

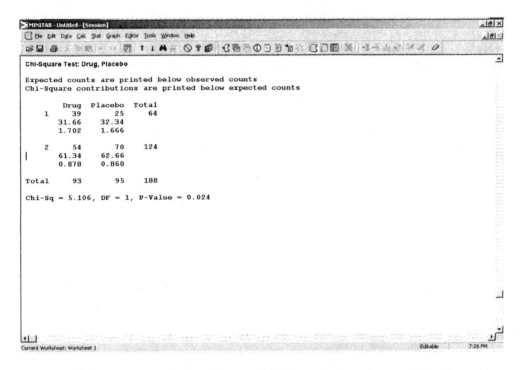

Notice that the test statistic is Chi-Sq = 5.106 and the P-value = .024. Since this P-value is smaller than α = .10, you should reject the null hypothesis. Thus, there is evidence to conclude that the drug should be used as part of the treatment.

Section 10.3

> ▶ **Example 3 (pg. 583)** Performing a Two-Sample F-Test

A restaurant manager designed a system to decrease the variance in customer wait times. To perform the two-sample F-test, click on **Stat → Basic Statistics → 2 Variances.** Click on **Summarized data**, since this example has only sample sizes and variances. Beside **First**, enter the Sample size and Variance for the old system and beside **Second**, enter the data for the new system. Click on **OK** to see the results of the F-test in both the Graph Window and in the Session Window. Below are the results from the Session Window.

Test for Equal Variances

```
95% Bonferroni confidence intervals for standard deviations

Sample   N    Lower   StDev    Upper
     1  10  13.0825     20   40.2731
     2  21  11.8017     16   24.4620


F-Test (Normal Distribution)
Test statistic = 1.56, p-value = 0.388
```

The F-test results show that the test statistic = 1.56 and the P-value = .388. Since this is a large P-value, you would fail to reject the null hypothesis. Thus, the two variances are approximately equal.

◀

Section 10.4

> **Example 2 (pg. 593)** ANOVA Tests

Open worksheet **Airline** found in the MINITAB folder **ch10.** The data for the three airlines should be in C1, C2, and C3. To perform a one-way analysis of variance, click on **Stat → ANOVA → One Way (Unstacked).** Select all three columns for **Responses (in separate columns)** and click on **OK.** The results of the test will be in the Session Window.

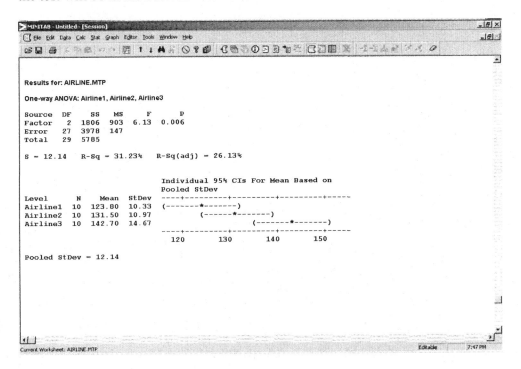

Notice that the test statistic is listed (F = 6.13), as well as the P-value (.006). Since the P-value = .006, and this is smaller than α = .01, you should reject the null hypothesis. Thus, there is a difference in the average flight times.

> **Exercise 5 (pg. 595)**   Costs per month of different toothpastes

Open worksheet **Ex10_4-5** found in the MINITAB folder **ch10.** The data for the three different degrees of abrasiveness should be in C1, C2, and C3. To perform a one-way analysis of variance, click on **Stat → ANOVA → One Way (Unstacked).** Select all three columns and click on **OK.** The results of the test will be in the Session Window.

One-way ANOVA: Very Good, Good, Fair

```
Source   DF     SS      MS      F      P
Factor    2   0.518   0.259   1.02   0.376
Error    26   6.629   0.255
Total    28   7.148

S = 0.5049    R-Sq = 7.25%    R-Sq(adj) = 0.12%

                                 Individual 95% CIs For Mean Based
                                 On Pooled StDev
Level        N    Mean   StDev   ---+---------+---------+---------+-
-----
Very Good   12  0.5358  0.3222      (-----------*-----------)
Good        12  0.8267  0.6657                  (-----------*--------
---)
Fair         5  0.6300  0.3913   (-----------------*----------------
--)
-----
                                 ---+---------+---------+---------+-
-----
                                 0.25      0.50      0.75      1.00

Pooled StDev = 0.5049
```

Since the P-value = .376 and is larger than α, you should not reject the null hypothesis. Thus, there is not enough evidence that the average cost per month is the different for the three types of toothpaste.

▶ **Exercise 9 (pg. 596)**   Days spent in a Hospital

Open worksheet **Ex10_4-9** found in the MINITAB folder **ch10.** The data for the
four different regions of the United States should be in C1 - C4. To perform a
one-way analysis of variance, click on **Stat → ANOVA → One-Way
(Unstacked).** Select all four columns and click on **OK.** The results of the test
will be in the Session Window.

One-way ANOVA: Northeast, Midwest, South, West

```
Source  DF       SS    MS     F       P
Factor   3     5.61  1.87  0.56   0.648
Error   29    97.30  3.36
Total   32   102.91

S = 1.832    R-Sq = 5.45%    R-Sq(adj) = 0.00%

                                 Individual 95% CIs For Mean Based on
                                 Pooled StDev
Level       N    Mean   StDev  --------+---------+---------+---------
+-
Northeast   9   5.444   2.351                (-----------*------------)
Midwest     9   4.444   1.667  (-----------*------------)
South       7   4.857   1.574    (--------------*-------------)
West        8   4.500   1.512  (-----------*------------)
                               --------+---------+---------+---------
                                  4.0       5.0       6.0

Pooled StDev = 1.832
```

Since the P-value = .648 and is larger than α, you should NOT reject the null
hypothesis. Thus, the average number of days spent in a hospital is the same for
all four regions of the United States.

◀

▶ Technology Lab (pg. 609) Teacher Salaries

Open worksheet **Tech10_a** found in the MINITAB folder **ch10.** The data for the teacher salaries in three states should be in C1 - C3.

1. Since the three states represent different populations and the three samples were randomly chosen, the samples are independent.

2. MINITAB has a test for normality. Click on **Stat → Basic Statistics → Normality Test.** Select C1 for the **Variable** and **Kolmogorov-Smirnov** for the **Test of Normality.** Click on **OK** and a normal plot will be displayed.

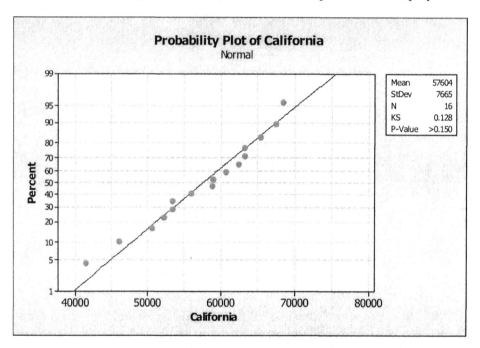

    Notice the P-value is listed with results. Since this P-value is larger than .10 (our α), then you can assume that the data is approximately normal. Repeat this test for the other two columns of data.

3. To test if the 3 samples have equal variances, MINITAB requires that the data be stacked into one column with a second column identifying which sample each data value came from. To do this, click on **Data → Stack → Columns.** Select all three columns to be stacked on top of each other. **Column of current worksheet** should be C4 and **Store subscripts in** C5. The subscripts will be numbers 1 or 2 or 3 to indicate which column the data value came from, or if you select **Use variable names in subscript column** then the column names will be your subscripts. Next perform the Test for

Equal Variances. To do this, click on **Stat → ANOVA → Test for Equal Variances.** On the input screen, C3 is the **Response** variable and C4 is the **Factor.** Enter an appropriate **Title** and click on **OK.** This test produces quite a lot of output, however you are only interested in the results of the F-test. You can see the results in both the Session Window and the Graph Window. Below are the results from the Graph Window. Bartlett's Test assumes normality of the data.

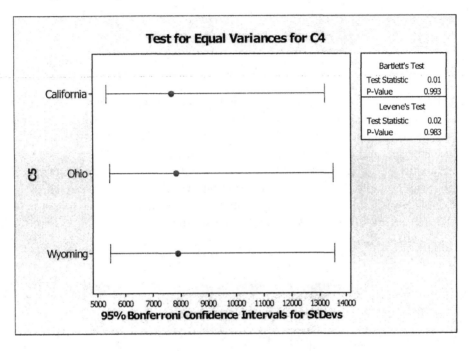

4. To do a one-way ANOVA, click on **Stat → ANOVA → One-Way (Unstacked).** Select all three columns and click on **OK.** The results of the test will be in the Session Window. Look at the P-value. If it is smaller than α, you should reject the null hypothesis.

5. Repeat Exercises 1 - 4 using worksheet **Tech10_b** found in the MINITAB folder **ch10.**

Nonparametric Tests

Section 11.1

> ▸ Example 3 (pg. 617) Using the Paired-Sample Sign Test

Open worksheet **Rehabilitation** which is found in the **ch11** MINITAB folder. The 'Before' data is in C1 and the 'After' data is in C2. Since we are interested in the difference between C1 and C2, we will create a new column that is C1 - C2. Click on **Calc → Calculator. Store result in variable** C3 and calculate the **Expression** C1 - C2. Click on **OK** and C3 should contain the differences. Now perform a 1-sample Sign test on C3. Click on **Stat → Nonparametrics → 1-sample Sign.** Select C3 as the **Variable.** Since you would like to test if the number of repeat offenders has decreased after the special course, you would expect that the differences in C3 would be greater than 0. Thus, enter 0 for **Test median** and select **greater than** for the **Alternative.**

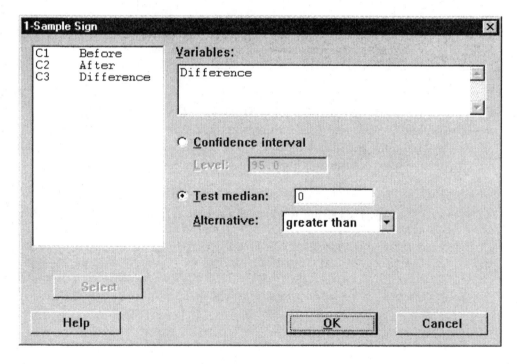

Click on **OK** and the results will be in the Session Window.

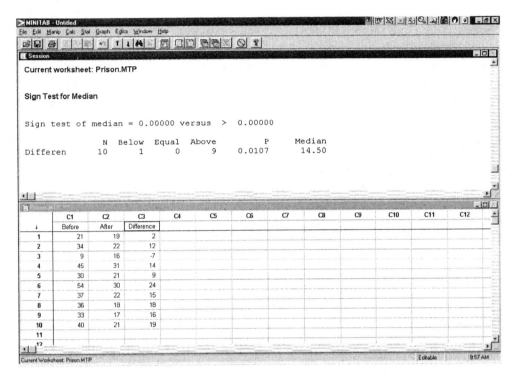

Notice that the P-value = .0107. Since this value is smaller than α=.025, you would reject the null hypothesis. Thus, there is sufficient evidence to conclude that the number of repeat offenders decreases after taking the special course.

> ▶ Exercise 3 (pg. 618)   Credit Card Charges

Enter the data into C1 in the MINITAB Data Window. (Do not type in the $ sign.) To perform the Sign Test, click on **Stat → Nonparametrics → 1-sample Sign.** Select C1 as the **Variable.** Since you would like to test if the median amount of new credit card charges was more than $300, enter 300 for **Test median** and select **greater than** for the **Alternative.**

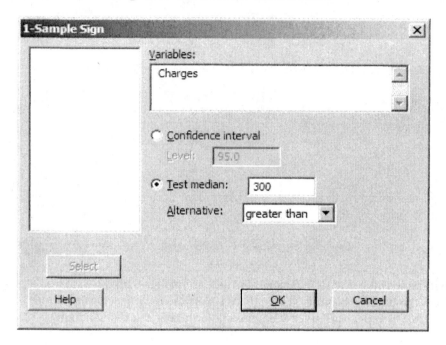

Click on **OK** and the results will be displayed in the Session Window.

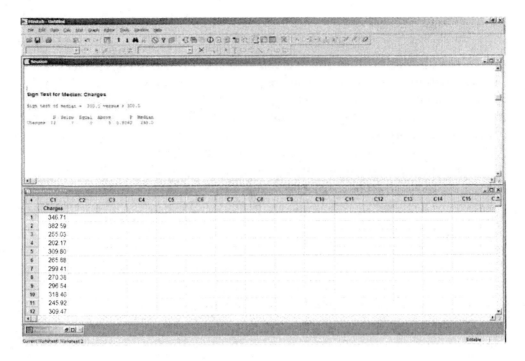

Notice that the P-value = .8062. Since this is such a large P-value, you would fail to reject the null hypothesis. Thus, the accountant can **not** conclude that the median amount of new charges was more than $300.

▶ Exercise 16 (pg. 620) Does therapy decrease pain intensity?

Open worksheet **Ex11_1-16** which is found in the **ch11** MINITAB folder. Pain intensity before taking anti-inflammatory drugs is in C1 of the MINITAB Data Window and Pain intensity after taking the drug is in C2. Calculate the differences. Click on **Calc → Calculator.** Next **Store result in variable** C3 and calculate the **Expression** C1-C2. Click on **OK** and C3 should contain the differences. Now perform a 1-sample Sign test on C3. Click on **Stat → Nonparametrics → 1-sample Sign.** Select C3 as the **Variable.** Since you would like to test if the intensity scores have decreased after the anti-inflammatory drugs, you would expect that the differences in C3 would be greater than 0. Thus, enter 0 for **Test median** and select **greater than** for the **Alternative.** Click on **OK.**

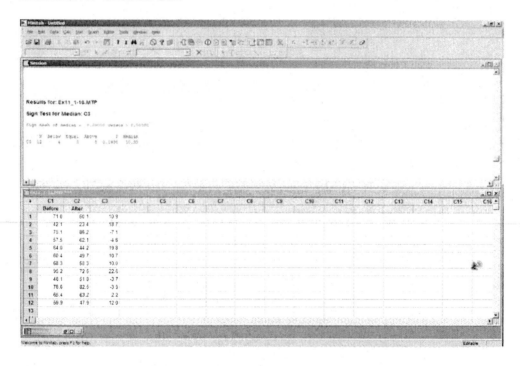

Notice that the P-value = .1938. Since this value is larger than α=.05, you would fail to reject the null hypothesis. Thus, there is not enough evidence to conclude that the pain intensity decreases with these drugs.

Section 11.2

▶ Example 1 (pg. 624) Performing a Wilcoxon Signed-Rank Test

Enter the 'With Music' data in C1 and 'Without Music' data in C2. To calculate the differences, click **Calc → Calculator.** **Store the result in variable** C3 and calculate the **Expression** C1-C2. Click on **OK** and C3 should contain the differences. To perform the Wilcoxon Signed Rank Test, click on **Stat → Nonparametrics → 1-sample Wilcoxon.** You should use C3 for the **Variable.** Since you are using the differences in this example, you want to compare the median difference to 0. So, enter 0 for **Test Median** and choose **not equal** as the **Alternative.**

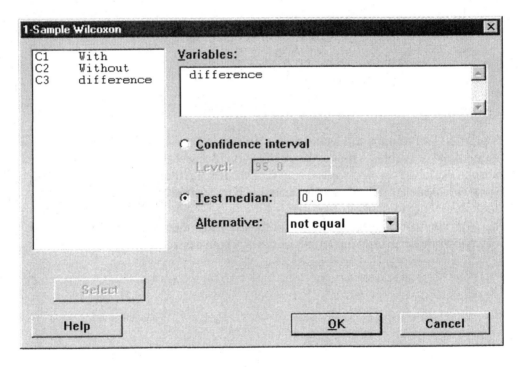

Click on **OK** to view the results of the test in the Session Window.

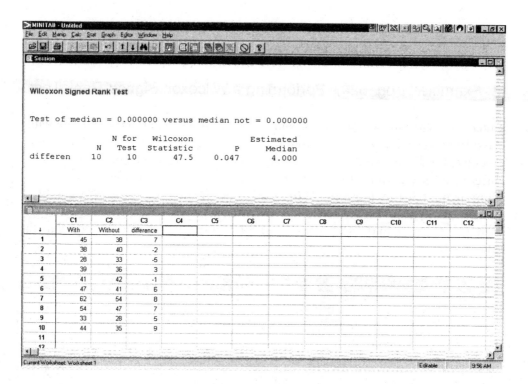

The MINITAB output tells you that the Wilcoxon Statistic is 47.5 and the P-value is .047. Although the textbook tells you to select the smaller of the absolute value of the two sums of the ranks, MINITAB simply uses the sum of the positive ranks. This makes no difference in interpreting the results. The important thing to notice is that the P-value = .047. Since this is smaller than α = .05, you would reject the null hypothesis. Thus, there is sufficient evidence to say that music affects the length of the workout sessions.

▸ **Example 2 (pg. 627)** Performing a Wilcoxon Rank Sum Test

Open worksheet **Earnings** which is found in the **ch11** MINITAB folder. Male
earnings are in C1 and Female earnings are in C2. In MINITAB, the Wilcoxon
Rank Sum Test is called the Mann-Whitney test. Click on **Stat →
Nonparametrics → Mann-Whitney.** Enter C1 for the **First Sample,** C2 for the
Second Sample, and select n**ot equal** as the **Alternative** since you want to see if
there is a difference between the earnings.

```
Mann-Whitney                                               ☒

   C1    Male Earning      First Sample:     'Male Earnings
   C2    Female Earni
                           Second Sample:    'Female Earnin


                           Confidence level:    95.0


                           Alternative:    not equal      ▼



         Select

     Help                         OK              Cancel
```

Click on **OK** and the results will be in the Session Window.

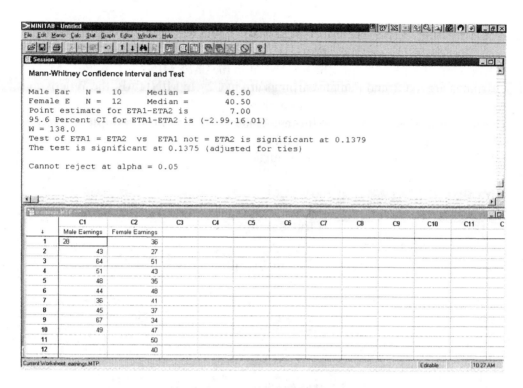

Look at the results carefully. The P-value is .1379. The rank sum for male earnings is also listed, W = 138. In this example, since the P-value is larger than α=.10, you would fail to reject the null hypothesis. Thus, there is no difference between male and female earnings.

▶ Exercise 7 (pg. 630) Teacher salaries

Open worksheet **Ex11_2-7** which is found in the **Chapter 11** MINITAB folder. Click on **Stat → Nonparametrics → Mann-Whitney.** Enter C1 for the **First Sample,** C2 for the **Second Sample**, and select **not equal** as the **Alternative** since you want to see if there is a difference between the salaries in Massachusetts and Connecticut. Click on **OK**. The results should be in the Session Window.

Results for: Ex11_2-7.MTP

Mann-Whitney Test and CI: Mass., Conn.

```
        N  Median
Mass.  12  51.500
Conn.  12  57.000


Point estimate for ETA1-ETA2 is -4.000
95.4 Percent CI for ETA1-ETA2 is (-8.001,0.001)
W = 117.0
Test of ETA1 = ETA2 vs ETA1 not = ETA2 is significant at 0.0606
The test is significant at 0.0596 (adjusted for ties)
```

Since the P-value = .0606 and is larger than α=.05, you would fail to reject the null hypothesis. There is not enough evidence to conclude there is a difference in teacher salaries in Massachusetts and Connecticut.

◀

Section 11.3

▶ **Example 1 (pg. 635)** Performing a Kruskal Wallis Test

Open worksheet **Actuaries** which is found in the **ch11** MINITAB folder. To perform a Kruskal-Wallis test, MINITAB requires that the data be stacked into one column with a second column identifying which sample each data value came from. To do this, click on **Data → Stack → Columns.** Select all three columns to be stacked on top of each other. Select **Column of current worksheet** and enter C4 and **Store subscripts in** C5. The subscripts will be numbers 1, 2, or 3 to indicate which column the data value came from. Be sure that **Use variable names in subscript column** is NOT selected.

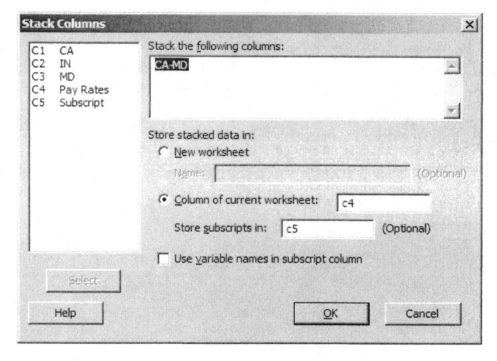

Click on **OK.** Name C4 Payrates and C5 State. Notice that in C5, 1 represents California, 2 represents Indiana, and 3 represents Maryland.

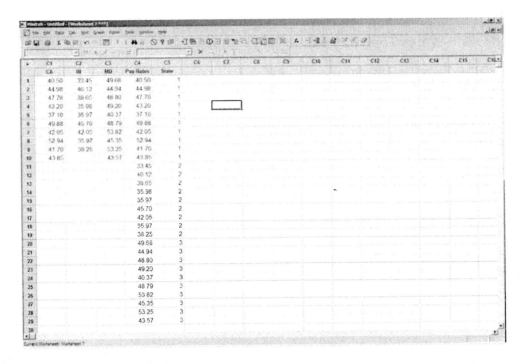

Now, to do the Kruskal-Wallis test, click on **Stat → Nonparametrics → Kruskal-Wallis.** The **Response** variable is Payrates (C4) and the **Factor** is State (C5).

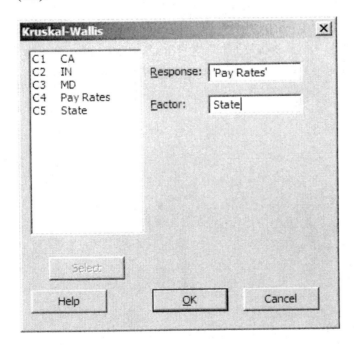

Click on **OK** and the results will be displayed in the Session Window.

Kruskal-Wallis Test: Pay Rates versus State

```
Kruskal-Wallis Test on Pay Rates

State     N   Median   Ave Rank       Z
1        10    43.53       16.1    0.48
2         9    38.25        7.1   -3.37
3        10    48.80       21.1    2.80
Overall  29                15.0

H = 13.12   DF = 2   P = 0.001
H = 13.13   DF = 2   P = 0.001   (adjusted for ties)
```

Notice that the test statistic is H=13.12 and the P-value=.001. With such a small P-value, you should reject the null hypothesis. So, there is a difference in the pay rates of the three states.

> **Exercise 3 (pg. 637)**   Are the insurance premiums different?

Open worksheet **Ex11_3-3** which is found in the **ch11** MINITAB folder. The data should be in C1, C2, and C3. To perform a Kruskal-Wallis test, MINITAB requires that the data be stacked into one column with a second column identifying which sample each data value came from. To do this, click on **Data → Stack → Columns.** Select all three columns to be stacked on top of each other. Select **Column of current worksheet** and enter C4 and **Store subscripts in** C5. The subscripts will be numbers 1, 2, or 3 to indicate which column the data value came from. Be sure that **Use variable names in subscript column** is NOT selected. Click on **OK.** Name C4 Premiums and C5 State. Notice that in C5, 1 represents Arizona, 2 represents Florida, and 3 represents Louisiana. Now, to do the Kruskal-Wallis test, click on **Stat → Nonparametrics → Kruskal-Wallis.** The **Response** variable is Premiums (C4) and the **Factor** is State (C5). Click on **OK** and the results will be displayed in the Session Window.

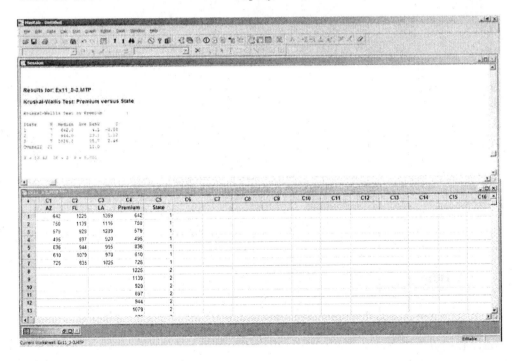

Notice that the test statistic is H=13.42 and the P-value=.001. With such a small P-value, you should reject the null hypothesis. So, there is a difference in the premiums of the three states.

Section 11.4

▶ Example 1 (pg. 640) The Spearman Rank Correlation
Coefficient

Enter the data into the MINITAB Data Window. Enter the Beef prices into C1
and the Lamb prices into C2. To rank the data values, click on **Data → Rank.**
On the input screen, you should **Rank data in** C1 and **Store ranks in** C3.
When you click on **OK**, the ranks of the Beef prices should be in C3.

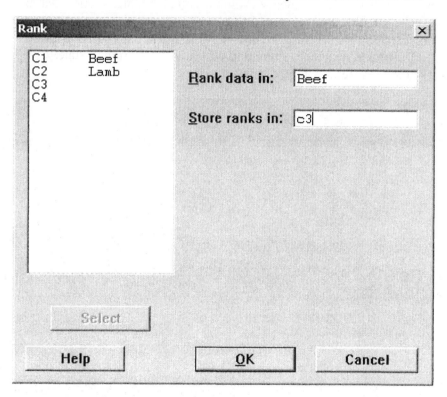

Repeat this using the Lamb prices and storing the ranks in C4. Now you should
have the ranks of the data in C3 and C4.

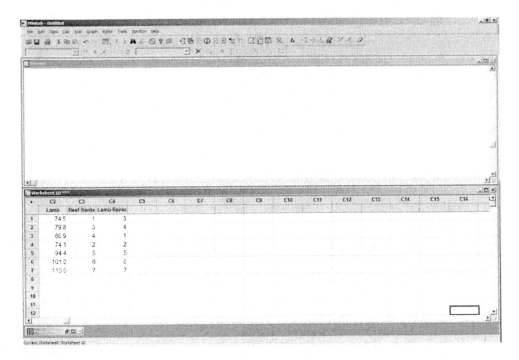

To calculate Spearman's Rank Correlation Coefficient, simply use Pearson's correlation on the ranks of the data. Click on **Stat → Basic Statistics → Correlation.** Enter C3 and C4 for the **Variables** and select **Display p-values.**

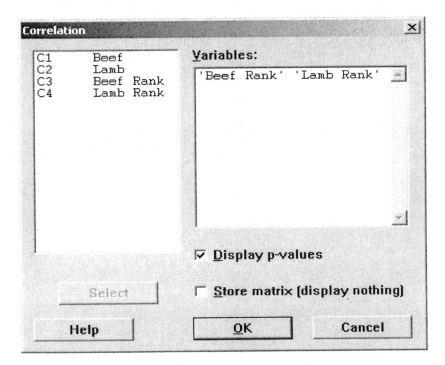

When you click on **OK**, the results will be displayed in the Session Window.

Correlations: Beef Ranks, Lamb Ranks

```
Pearson correlation of Beef Ranks and Lamb Ranks = 0.750
P-Value = 0.052
```

In this example, notice that the Spearman's Rank Correlation Coefficient is 0.75 and the P-value is .052. Since this P-value is larger than $\alpha = .05$, you would fail to reject the null hypothesis. Thus, you can NOT conclude that there is a significant correlation between beef and lamb prices from 1999 to 2005.

▶ Exercise 6 (pg. 643)   Is Air Conditioner Performance related to Price?

Open worksheet **Ex11_4-6** which is found in the **ch11** MINITAB folder. The overall score is in C1 and the price is in C2. First, rank the data. Click on **Data → Rank.** On the input screen, you should **Rank data in** C1 and **Store ranks in** C3. When you click on **OK**, the ranks of the Overall Scores should be in C3. Repeat this for the Prices and **store ranks in** C4. Now, calculate the correlation coefficient of the ranks. Click on **Stat → Basic Statistics → Correlation.** Enter C3 and C4 for the **Variables** and select **Display p-values.**

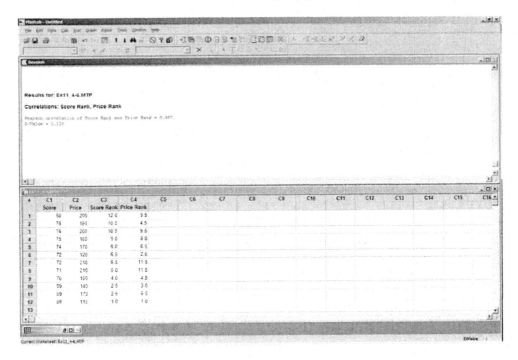

Notice that the Correlation Coefficient is .467 and the P-value is .126. Since the P-value is larger than α=.10, you would fail to reject the null hypothesis. Thus, there is not a significant correlation between Overall Score and Price.

Section 11.5

▶ **Example 3 (pg. 649)** The Runs Test

Enter the data into C1 of the MINITAB Data Window. MINITAB requires numeric data for the Runs Test, so you will have to code the data. Click on **Data → Code → Text to Numeric.** On the input screen, you should **Code data from columns** C1 and **into** C2. Code **Original values** "M" as a **New** value of 1 and "F" as 2.

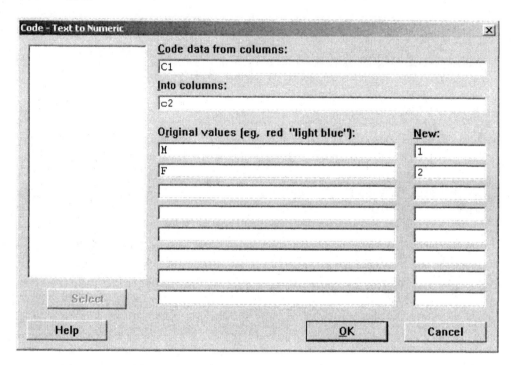

When you click on **OK**, the coded data should be in C2.

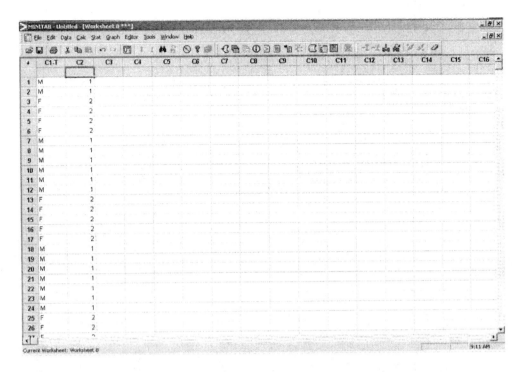

Next, click on **Stat → Nonparametrics → Runs Test.** Select C2 for the **Variable,** and select **Above and below mean.**

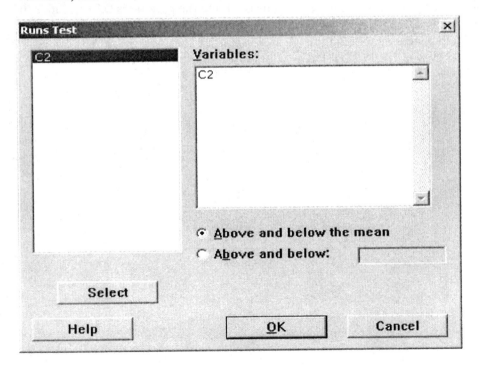

Click on **OK** and the results will be displayed in the Session Window.

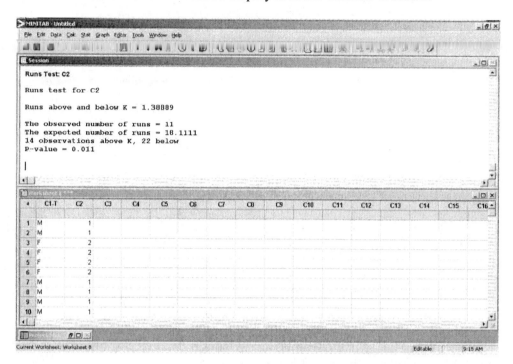

Notice that the observed number of runs is 11 and the p-value is 0.011. Since the p-value is less than .05, you can conclude that the selection of employees with respect to gender is not random.

> ▶ Technology Lab (pg. 661) Annual Income

Open worksheet **Tech11_a** which is found in the **ch11** MINITAB folder. The data should be in C1 - C4.

1. Construct a boxplot for all four regions. Click on **Graph → Boxplot → Multiple Y's Simple.** Select C1, C2, C3, and C4 for the **Graph variables.** Click on the **Labels** button and enter an appropriate title. Click on **OK** twice to view the boxplots.

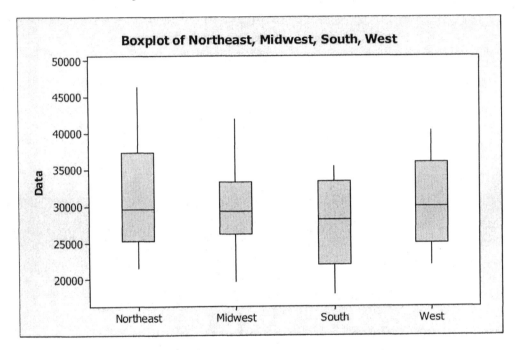

2. To perform a sign test on the data from the Midwest, click on **Stat → Nonparametrics → 1-Sample Sign.** Select South (C2) for the **Variable,** and enter 25000 for the **Test Median.** The **Alternative** should be **greater than** since the claim is that the median income in the Midwest is at least $25,000. Click on **OK** and the results will be displayed in the Session Window.

3. Recall that the Wilcoxon rank sum test is the same as the Mann-Whitney test in MINITAB. Click on **Stat → Nonparametrics → Mann-Whitney.** Enter Northeast (C1) for the **First sample** and South (C3) for the **Second sample.** The **Alternative** should be **not equal** since you are testing whether or not the median annual incomes are the same. Click on **OK** and the results will be displayed in the Session Window.

The TI-83 / TI-84 Graphing Calculator Manual

Kathleen McLaughlin
Manchester Community Technical College

Dorothy Wakefield
University of Connecticut

FOURTH EDITION

Elementary Statistics

Picturing *the* World

Larson | Farber

PEARSON

Prentice Hall

Upper Saddle River, NJ 07458

▶ Introduction

The TI-83/TI-84 Graphing Calculator Manual is one of a series of companion technology manuals that provide hands-on technology assistance to users of Larson/Farber *Elementary Statistics: Picturing the World, 4th Edition.*

Detailed instructions for working selected examples, exercises, and Technology sections from *Elementary Statistics: Picturing the World, 4th Ed.* are provided in this manual. To make the correlation with the text as seamless as possible, the table of contents includes page references for both the Larson/Farber text and this manual.

▶ Contents:

Getting Started with the TI-83 and TI-84 Graphing Calculators

▶ Overview

This manual is designed to be used with the TI-83 and TI-84 families of Graphing Calculators. These calculators have a variety of useful functions for doing statistical calculations and for creating statistical plots. The commands for using the statistical functions are basically the same for the TI-83's and TI-84's. All TI-84 calculators, the TI-83 Plus Calculator and the TI-83 Silver Edition can receive a variety of software applications that are available through the TI website (www.ti.com). TI also will provide downloadable updates to the operating systems of these calculators. These features are not available on the TI-83.

Your textbook comes with data files on the CD data disk that can be loaded onto any of the TI-83 or TI-84 calculators. The requirements for the transfer of data differ from one calculator to another. Some of these calculators (TI-84's and TI-83 Silver Edition) are sold with the necessary connection. For the TI-83 or TI-83 Plus, you can purchase a Graph Link manufactured by Texas Instruments which connects the calculator to the computer. (Note: In order to do examples in this manual, you can simply enter the data values for each example directly into your calculator. It is not necessary to use the graph link to download the data into your calculator. The download procedure using the computer link is an optional way of entering data.)

Throughout this manual all instructions and screen shots use the TI-84. These instructions and screen shots are also compatible with the TI-83 calculators.

Before you begin using the TI-83 or TI-84 calculator, spend a few minutes becoming familiar with its basic operations. First, notice the different colored keys on the calculator. On the TI-84's, the white keys are the number keys; the light gray keys on the right are the basic mathematical functions; the dark gray keys on the left are additional mathematical functions; the remaining dark gray keys are the advanced functions; the light gray keys just below the viewing screen are used to set up and display graphs, and the light gray arrow keys are used for moving the cursor around the viewing screen. On the TI-83's, the white keys are the number keys; the blue keys on the right are the basic mathematical functions; the dark gray keys on the left are additional mathematical functions;

the remaining dark gray keys are the advanced functions; the blue keys just below the viewing screen are used to set up and display graphs, and the blue arrow keys are used for moving the cursor around the viewing screen.

The primary function of each key is printed in white on the key. For example, when you press **STAT**, the STAT MENU is displayed.

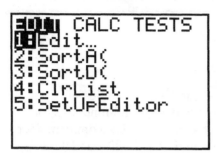

The secondary function of each key is printed in blue on the TI-84's (yellow on the TI-83's) above the key. When you press the **2ⁿᵈ** key (found in the upper left corner of the keys), the function printed above the key becomes active and the cursor changes from a solid rectangle to an ⬆ (up-arrow). For example, when you press **2ⁿᵈ** and the $\boxed{x^2}$ key, the $\sqrt{}$ function is activated.

The notation used in this manual to indicate a secondary function is '**2ⁿᵈ**' followed by the name of the secondary function. For example, to use the LIST function, found above the **STAT** key, the notation used in this manual is **2ⁿᵈ [LIST]**. The LIST MENU will then be activated and displayed on the screen.

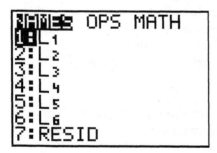

The alpha function of each key is printed in green above the key. When you press the green **ALPHA** key, the function printed in green above the key is activated and the cursor changes from a solid rectangle to **A**.

In this manual you will find detailed explanations of the different statistical functions that are programmed into the TI-83 and TI-84 graphing calculators. These explanations will accompany selected examples from your textbook. This will give you the opportunity to learn the various calculator functions as they apply to the specific statistical material in each chapter.

▸ Getting Started

To operate the calculator, press **ON** in the lower left corner of the calculator. Begin each example with a blank screen, with a rectangular cursor flashing in the upper left corner. If you turn on your calculator and you do not have a blank screen, press the **CLEAR** key. You may have to press **CLEAR** a second time in order to clear the screen. If using the **CLEAR** key does not clear the screen, you can push **2ⁿᵈ [QUIT]** (Note: **QUIT** is found above the **MODE** key.)

▸ Helpful Hints

To adjust the display contrast, push and release the **2ⁿᵈ** key. Then push and hold the up arrow ▲ to darken or the down arrow ▼ to lighten.

The calculator has an automatic turn off that will turn the calculator off if it has been idle for several minutes. To restart, simply press the **ON** key.

There are several different graphing techniques available on the TI-83 and TI-84 calculators. If you inadvertently leave a graph on and attempt to use a different graphing function, your graph display may be cluttered with extraneous graphs, or you may get an ERROR message on the screen.

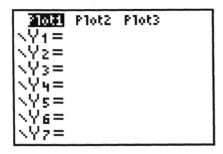

There are several items that you should check before graphing anything. First, press the **Y=** key, found in the upper left corner of the key pad, and clear all the Y-variables. The screen should look like the following display:

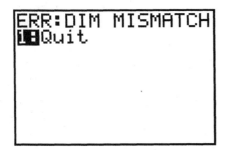

If there are any functions stored in the Y-variables, simply move the cursor to the line that contains a function and press **CLEAR** **ENTER**.

Next, press **2ⁿᵈ** **[STAT PLOT]** (found on the **Y=** key) and check to make sure that all the STAT PLOTS are turned **OFF**.

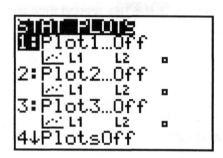

If you notice that a Plot is turned **ON**, select the Plot by using the down arrow key to highlight the number to the left of the Plot , press **ENTER** and move the cursor to **OFF** and press **ENTER**. Press **2ⁿᵈ** **[QUIT]** to return to the home screen.

Now you are ready to get started with your calculator.   Enjoy!!

Introduction to Statistics

CHAPTER

1

▶ Technology (pg. 36 - 37)   Generating Random Numbers

The first step is to initialize your calculator to generate random numbers by setting a unique starting value called a *seed*. You set the *seed* by selecting any 'starting number' and storing this number in **rand**. Suppose, for this example, that we select the number '34' as the starting number. Type **34** into your calculator and press the **STO** key found in the lower left section of the calculator keys. Next press the **MATH** key found in the upper left section of the calculator keys. The Math Menu will appear.

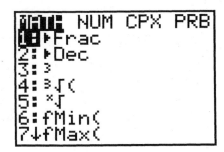

Use the right arrow key, 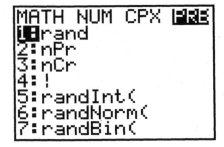 found in the upper right section of the calculator keys, to move the cursor to highlight **PRB**. The Probability Menu will appear.

The first selection on the **PRB** menu is **rand,** which stands for 'random number'. Notice that this highlighted. Simply press **ENTER** twice and the starting value

of '34' will be stored into **rand** and will be used as the *seed* for generating
random numbers. (Note: This example uses '34' as the *seed,* but you should use
your own number as a seed for your random number generator. You only need to
do this initialization process once. You do not need to do this every time you
generate random numbers.)

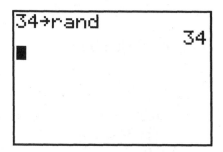

Now you are ready to generate random numbers.
To generate a random sample of integers, press **MATH** and the Math Menu will
appear.

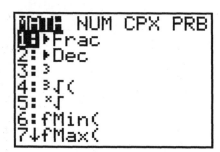

Use the right arrow key, ▶ , to move the cursor to highlight **PRB**. The
Probability Menu will appear.

MATH NUM CPX **PRB**
1▪rand
2:nPr
3:nCr
4: !
5:randInt(
6:randNorm(
7:randBin(

Select **5:RandInt(** by using the down arrow key, ▼ , to highlight it and
pressing **ENTER** or by pressing the 5 key. **RandInt(** should appear on the

screen. This function requires three values: the starting integer, followed by a comma (the comma is found on the black key above the [7] key), the ending integer, followed by a comma and the number of values you want to generate. Close the parentheses and press **ENTER**. (Note: It is optional to close the parenthesis at the end of the command.)

For an example, suppose you want to generate 15 values from the integers ranging from 1 to 50. The command is **randInt(1,50,15)**.

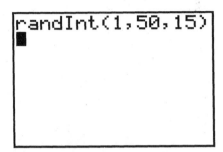

Press **ENTER** and a partial display of the 15 random integers should appear on your screen. (Note: your numbers will be different from the ones you see here.)

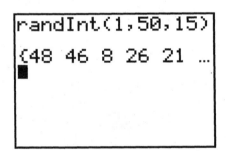

Use the right arrow to scroll through your 15 items. You might find that you have some duplicate values. The TI-83/84 uses a method called "sampling with replacement" to generate random numbers. This means that it is possible to select the same integer twice.

In the example in the text, you are asked to select a random sample of 15 cars from the 167 cars that are assembled at an auto plant. One way to choose the sample is to number the cars from 1 to 167 and randomly select 15 different cars. This sampling process is to be "without replacement." Since the TI-83/84 samples "with replacement", the best way to obtain 15 different cars is to generate more than 15 random integers, and to discard any duplicates. To be safe, you should generate 20 random integers.

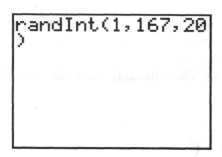

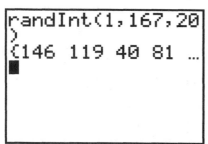

Use the right arrow to scroll to the right to see the rest of the list and write down the first 15 distinct values.

Exercises

1. You would like to sample 10 distinct brokers. The TI-83/84 samples "with replacement" so you may have duplicates in your sample. To obtain 10 distinct values, try selecting 15 random numbers. Press **MATH**, highlight **PRB** and select **5:RandInt(**. Press **ENTER** and type in **1,86,15).** Press **ENTER** and the random integers will appear on the screen. Use the right arrow to scroll through the output and choose the first ten distinct integers. (Note: If you don't have 10 distinct integers, simply press **ENTER** to generate 15 more integers and pick as many new integers as you need from this group to complete your sample of 10).

2. You would like to sample 25 distinct camera phones. To obtain 25 distinct values, try selecting 30 random numbers. Press **MATH**, highlight **PRB** and select **5:RandInt(**. Press **ENTER** and type in **1,300,30).** Press **ENTER** and the random integers will appear on the screen. Use the right arrow to scroll through the output and choose the first twenty-five distinct integers. (Note: If you don't have 25 distinct integers, simply press **ENTER** to generate 30 more integers and pick as many new integers as you need from this group to complete your sample of 25).

3. This example does not require distinct digits, so duplicates are allowed. You can select three random samples of size n=5 by using **randInt** to create each sample. To generate the first sample, press **MATH**, highlight **PRB** and select **5:RandInt(** by scrolling down through the list and highlighting **5:RandInt(** and pressing **ENTER** or by highlighting **PRB** and pressing **5**. Type in the starting digit: 0; the ending digit: 9, and the number of selections: 5.

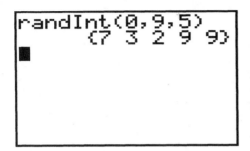

Add the digits and divide by 5 to get the average for the sample.

To generate the second and third samples, simply press **ENTER** twice. Calculate the sample average for the 2nd and 3rd sample.

To calculate the population average, add the digits 0 through 9 and divide the answer by 10.

Compare your three sample averages with your population average.

4. Use the same procedure as in Exercise 3 with a starting value of 0 and an ending value of 40 and a sample size of 7.

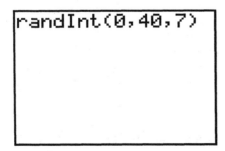

5. Use **randInt(1,6,60).** Store the results in L1 by pressing **STO** **2nd** **[L1]** **ENTER**. (Note: **[L1]** is found above the **1** key.) Then sort L1 by pressing **2nd** **[LIST]**, (**[LIST]** is found above the **STAT** key), highlight **OPS** and

select **1:SortA(** and press 2nd **[L1]** **ENTER**. Press 2nd **[L1]** **ENTER** and a partial list of the integers in L1 will appear on the screen. Use the right arrow to scroll through the list and count the number of 1's, 2's, 3's, ... and 6's. Record the results in a table.

7. Use **randInt(0,1,100).** Store the results in L1. Press 2nd **[LIST]** , highlight **MATH** and select **5:sum(** and press 2nd **[L1]**. Close the parentheses and press **ENTER**.

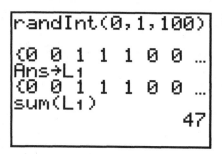

The number in your output is the sum of L1. Since L1 consists of 0's and 1's, the sum is actually the number of tails in your random sample. The (number of heads) = 100 - (no. of tails).

Descriptive Statistics

Section 2.1

▸ Example 7 (pg. 48) Constructing a histogram for a frequency
distribution

To create this histogram, you must enter information into List1 (**L1**) and List 2
(**L2**). Refer to the frequency distribution on pg. 43 in your textbook. You will
enter the midpoints into **L1** and the frequencies into **L2**. To enter the data, press
STAT and the Statistics Menu will appear. Notice that **EDIT** is highlighted. The
first selection in the **EDIT** menu, **1:Edit**, is also highlighted.

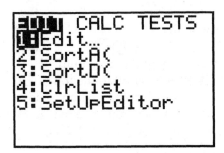

Press ENTER and lists **L1**, **L2** and **L3** will appear.

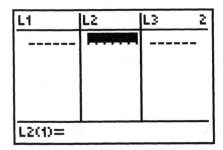

If the lists already contain data, you should clear them before beginning this
example. To clear data from a list, move your cursor so that the List name (**L1**,
L2, or **L3**) of the list that contains data is highlighted.

```
 L1       L2      L3      1
  1        4     ------
  2        6
  3        8
 10
------

L1 ={1,2,3,10}
```

Press **CLEAR** **ENTER**. Repeat this process until all three lists are empty.

```
L1       L2      L3      1
         4      ------
         6
         8
         ------


L1(1)=
```

To enter the midpoints into **L1,** move your cursor so that it is positioned in the 1st position in **L1**. Type in the first midpoint, **12.5,** and press **ENTER** or use the **Down Arrow**. Enter the next midpoint, **24.5.** Continue this process until all 7 midpoints are entered into **L1**. Now use the **Up Arrow** to scroll to the top of **L1**. As you scroll through the data, check it. If a data point is incorrect, simply move the cursor to highlight it and type in the correct value. When you have moved to the 1st value in **L1**, use the right arrow to move to the first position in **L2**. Enter the frequencies into **L2**.

```
L1       L2      L3      2
12.5     6      ------
24.5    10
36.5    13
48.5     8
60.5     5
72.5     6
84.5     2

L2(1)=6
```

To prepare to construct the histogram, press the **Y=** key and clear all the Y-registers. To graph the histogram, press **2nd** [STAT PLOT] (located above the **Y=** key).

Select Plot1 by pressing **ENTER**.

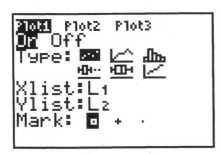

Notice that Plot1 is highlighted. On the next line, notice that the cursor is flashing on **ON** or **OFF**. Position the cursor on **ON** and press **ENTER** to select it. The next two lines on the screen show the different types of graphs. Move your cursor to the symbol for histogram (3rd item in the 1st line of **Type**) and press **ENTER**.

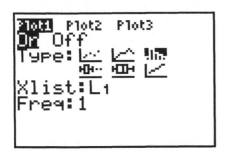

The next line is **Xlist**. Use the **Down Arrow** to move to this line. On this line, you tell the calculator where the data (the midpoints) are stored. In most graphing situations, the data are entered into **L1** so **L1** is the default option. Notice that the cursor is flashing on **L1**. Push **ENTER** to select **L1**. The last line is the frequency line. On this line, **1** is the default. The cursor should be flashing on **1**. Change **1** to **L2** by pressing 2nd **[L2]**.

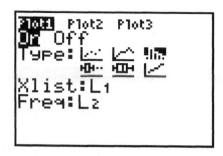

To view a histogram of the data, press **ZOOM**.

There are several options in the Zoom Menu. Using the **Down Arrow**, scroll down to option 9, **ZoomStat,** and press **ENTER**. A histogram should appear on the screen.

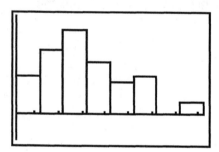

This histogram is not exactly the same as the one on pg. 48 of your textbook. You can adjust the histogram so that it does look exactly like the one in your text. Press **Window** and set **Xmin** to 12.5, **Xmax** to 96.5 (this one extra midpoint is needed to complete the picture), and **Xscl** equal to 12, which is the difference between successive midpoints in the frequency distribution. Note: It is not necessary for you to make any changes to **Ymin**, **Ymax** or **Yscl**.

```
WINDOW
 Xmin=12.5
 Xmax=96.5
 Xscl=12
 Ymin=-3.90897
 Ymax=15.21
 Yscl=.1
 Xres=1
```

Press **GRAPH**.

Notice the **TRACE** key (located just below the output screen.) If you press it, a flashing cursor, *, will appear at the top of the 1ˢᵗ bar of the histogram.

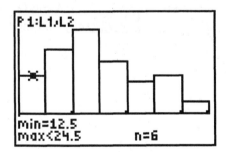

The value of the smallest midpoint is displayed as 12.5 and the number of data points in that bar is displayed as n = 6. Use the **Right Arrow** to move through each of the bars.

Now that you have completed this example, turn Plot1 **OFF**. Using **2ⁿᵈ** [STAT PLOT], select Plot1 by pressing **ENTER** and highlighting **OFF**. Press **ENTER** and **2ⁿᵈ** [QUIT]. (Note: Turning Plot1 **OFF** is optional. You can leave it ON but leaving it ON will affect other graphing operations.)

▶ **Exercise 29 (pg. 52)** Construct a frequency histogram using 6 classes

Push **STAT** and **ENTER** to select **1:Edit**. Highlight **L1** at the top of the first list and press **CLEAR** and **ENTER** to clear the data in **L1**. You can also clear **L2** although you will not be using **L2** in this example. Enter the data into **L1**. Scroll through your completed list and verify each entry.

```
L1      L2      L3      1
2114    ------  ------
2468
7119
1876
4105
3183
1932
L1(1)=2114
```

To prepare to construct the histogram, press the **Y=** key and clear all the Y-registers. To set up the histogram, push **2nd** [STAT PLOT] and **ENTER** to select **Plot 1**. Turn ON **Plot 1**, set **Type** to **Histogram**, set **Xlist** to **L1**. In this example, you must set **Freq** to **1**. If the frequency is set on **L2** move the cursor so that it is flashing on **L2** and press **CLEAR**. The cursor is now in ALPHA mode (notice that there is an "A" flashing in the cursor). Push the **ALPHA** key and the cursor should return to a solid flashing square. Type in the number **1.**

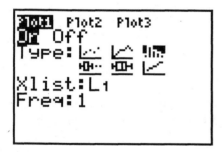

Press **ZOOM** and scroll down to **9:ZoomStat** and press **ENTER** and a histogram will appear on the screen. Press **Window** to adjust the Graph Window. Set **Xmin** equal to 1000 (the smallest data value) and **Xmax** equal to 7120 (a value, rounded to the nearest ten, that is larger than the largest data point in the dataset). To set the scale so that you will have 6 classes, calculate (**Xmax** - **Xmin**)/6: (7120 – 1000)/6 = 1020. Use this for **Xscl**. (Note: You do not need to change the values for **Ymin**, **Ymax** or **Yscl**.)

```
WINDOW
 Xmin=1000
 Xmax=7120
 Xscl=1020
 Ymin=-3.90897
 Ymax=15.21
 Yscl=.1
 Xres=1
```

Press **GRAPH** and the histogram should appear.

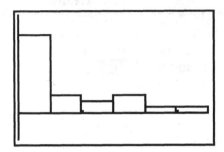

You can press **TRACE** and scroll through the bars of the histogram.
Min and Max values for each bar will appear along with the number of data
points in each class.

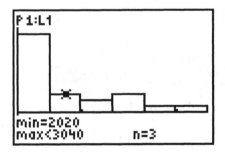

Notice, for example, with the cursor highlighting the second bar of the histogram,
you will see that **n=3** appears, indicating that there are 3 data points in the second
class which contains values from 2020 to 3039.

Note: After completing a graph, you should turn the graph **OFF**. Using 2nd
[STAT PLOT], select Plot1 by pressing **ENTER** and highlighting **OFF**. Press
ENTER and 2nd [QUIT]. (Note: Turning Plot1 **OFF** is optional. You can leave
it ON but
leaving it ON will affect other graphing operations.)

Section 2.2

▶ **Example 5 (pg. 59)** Constructing a Pareto Chart

To construct a Pareto chart to represent the causes of inventory shrinkage, you
must enter numerical labels for the specific causes into **L1** and the corresponding
costs into **L2**. Since the bars are positioned in descending order in a Pareto chart,
the labels in **L1** will represent the causes of inventory shrinkage from the most
costly to the least costly. Press **STAT** and press **ENTER** to select **1:Edit.**
Highlight the name "**L1**" and press **CLEAR** and **ENTER**. Enter the numbers
1,2,3 and 4 into **L1**. These numbers represent the four causes of inventory
shrinkage: 1 = employee theft, 2 = shoplifting, 3 = administrative error and 4 =
vendor fraud. Move your cursor to highlight "**L2**" and press **CLEAR** and
ENTER Enter the corresponding costs: 15.6, 14.7, 7.8 and 2.9.

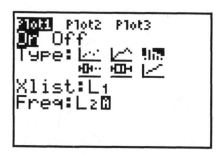

To prepare to construct the Pareto chart, press the **Y=** key and clear all the Y-
registers. To draw the Pareto chart, press **2ⁿᵈ** [STAT PLOT]. Press **ENTER**
and set up **Plot 1**. Highlight **On** and press **ENTER**. Highlight the histogram icon
for **Type** and press **ENTER.** Set **Xlist** to **L1** and **Freq** to **L2**.

Press **ZOOM** and scroll down to **9:ZoomStat** and press **ENTER** and a
histogram will appear on the screen. To create the Pareto chart with unconnected
bars, press **Window** and set Xmin = 1 (smallest value in **L1**), Xmax = 5 (one
more than the largest value in **L1**) and set Xscl = 0.5. (Note: You do not need to

change the values for **Ymin**, **Ymax** or **Yscl**.) Press **GRAPH** to view the Pareto Chart.

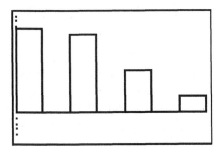

If you press **TRACE**, notice, for the first bar, min=1, max <1.5 and n = 15.6. The first bar represents the No. 1 cause of inventory shrinkage. N = 15.6 is the frequency (in millions of dollars) for cause #1. You can use the Right Arrow key to scroll through the remaining bars.

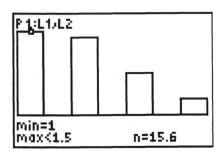

Note: After completing a graph, you should turn the graph **OFF**. Using **2**[nd] [STAT PLOT], select Plot1 by pressing **ENTER** and highlighting **OFF**. Press **ENTER** and **2**[nd] [QUIT]. (Note: Turning Plot1 **OFF** is optional. You can leave it ON but leaving it ON will affect other graphing operations.)

▶ Try It Yourself 6 (pg. 60) Constructing a Scatterplot

Press **STAT** and select **1:Edit** from the **Edit menu**. Clear the lists and enter the data points for "Length of employment" into **L1** and the data points for "Salary" into **L2**.

To prepare to construct the scatterplot, press the **Y=** key and clear all the Y-registers. To construct the scatterplot, press **2ⁿᵈ** [STAT PLOT] and select **1:Plot 1** and **ENTER**. Turn ON **Plot 1**. Set the **Type** to **Scatterplot** which is the 1ˢᵗ icon in the **Type** choices. For **Xlist** select **L1** and for **Ylist** select **L2**. For **Marks** use the first choice, the small square. Press **ZOOM** and press **9** for **ZoomStat** and view the graph.

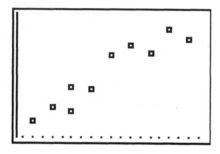

As you can see from the scatterplot, as the length of employment increases, the salary tends to increase.

Note: After completing a graph, you should turn the graph **OFF**. Using **2ⁿᵈ** [STAT PLOT], select Plot1 by pressing **ENTER** and highlighting **OFF**. Press **ENTER** and **2ⁿᵈ** [QUIT]. (Note: Turning Plot1 **OFF** is optional. You can leave it ON but leaving it ON will affect other graphing operations.)

▶ Example 7 (pg. 61) Constructing a Time Series Chart

Press **STAT** and select **1:Edit** from the **Edit Menu**. Clear **L1** and **L2**. Enter the "years" into **L1** and the "subscribers" into **L2**.

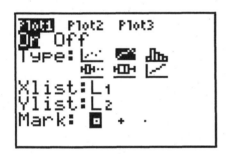

To prepare to construct the time series chart, press the **Y=** key and clear all the Y-registers. To construct the time series chart, press **2nd** [STAT PLOT] and select **1:Plot 1** and **ENTER**. Turn ON **Plot 1**. Set the **Type** to **Connected Scatterplot,** which is the 2nd icon in the **Type** choices. For **Xlist** select **L1** and for **Ylist** select **L2**. Next, there are three different types of **Marks** that you can select for the graph. The first choice, a small square, is the best one to use.

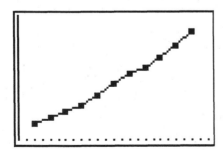

Press **ZOOM** and scroll down to **9:ZoomStat** and press **ENTER** or simply press **9** and **ZoomStat** will automatically be selected. The graph should appear on the screen.

Use **TRACE** and the Right Arrow key to scroll through the data values for each year. Notice for example, the number of subscribers is 69.2 million in 1998.

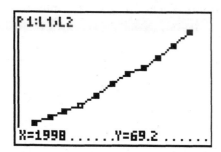

Note: After completing a graph, you should turn the graph **OFF**. Using 2nd [STAT PLOT], select Plot1 by pressing **ENTER** and highlighting **OFF**. Press **ENTER** and 2nd [QUIT]. (Note: Turning Plot1 **OFF** is optional. You can leave it ON but leaving it ON will affect other graphing operations.)

▶ Exercise 26 (pg. 64) Pareto Chart

To prepare to construct the Pareto chart, press the **Y=** key and clear all the Y-registers. Press **STAT** and select **1:Edit** from the **Edit Menu**. Clear the lists and enter the numbers 1 through 5 into **L1**. These numbers are labels for the five cities. (Note: 1= Miami, the city with the highest ultraviolet index, 2 = Atlanta, the city with the second highest ultraviolet index, etc.). Enter the ultraviolet indices in descending order into **L2**. Press **2ⁿᵈ** [STAT PLOT] and select **Plot 1** and press **ENTER**. Set the **Type** to **Histogram**. Set **Xlist** to **L1** and Freq to **L2**. Press **ZOOM** and scroll down to **9:ZoomStat** and press **ENTER** and a histogram will appear on the screen.

To create the Pareto chart with unconnected bars, press **Window** and set Xmin = 1 (smallest value in **L1**), Xmax = 6 (1 more than the largest value in **L1**) and set Xscl = 0.5. (Note: You do not need to change the values for **Ymin**, **Ymax** or **Yscl**.) Press **GRAPH** to view the Pareto Chart.

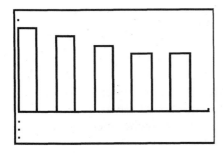

Note: After completing a graph, you should turn the graph **OFF**. Using **2ⁿᵈ** [STAT PLOT], select Plot1 by pressing **ENTER** and highlighting **OFF**. Press **ENTER** and **2ⁿᵈ** [QUIT]. (Note: Turning Plot1 **OFF** is optional. You can leave it ON but leaving it ON will affect other graphing operations.) ◀

▶ Exercise 28 (pg. 65) Scatterplot

To prepare to construct the scatterplot, press the **Y=** key and clear all the Y-registers. Press **STAT** and select **1:Edit** from the **Edit menu**. Clear the lists and enter the data points for "number of students per teacher" into **L1** and the data points for "average Teacher's salary" into **L2**.

To construct the scatterplot, press **2ⁿᵈ [STAT PLOT]** and select **1:Plot 1** and **ENTER**. Turn ON **Plot 1**. Set the **Type** to **Scatterplot** which is the 1ˢᵗ icon in the **Type** choices. For **Xlist** select **L1** and for **Ylist** select **L2**. For **Marks** use the first choice, the small square. Press **ZOOM** and press **9** for **ZoomStat** and view the graph.

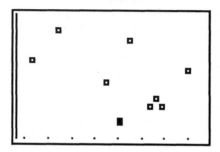

As you can see from the scatterplot, this data set has a large amount of scatter. You might notice a slight downward trend that suggests that the average teacher's salary decreases as the number of students per teacher increases.

Note: After completing a graph, you should turn the graph **OFF**. Using **2ⁿᵈ [STAT PLOT]**, select Plot1 by pressing **ENTER** and highlighting **OFF**. Press **ENTER** and **2ⁿᵈ [QUIT]**. (Note: Turning Plot1 **OFF** is optional. You can leave it ON but leaving it ON will affect other graphing operations.)

▸ Exercise 31 (pg. 65) Time Series Chart

To prepare to construct the Time Series chart, press the **Y=** key and clear all the Y-registers. Press **STAT** and select **1:Edit**. Enter the "Years" into **L1** and the "prices" into **L2**. Press **2ⁿᵈ** [STAT PLOT] and select 1: **Plot 1** and press **ENTER**. Turn **ON Plot 1** and select the **connected scatterplot** (2ⁿᵈ icon) as **Type**. Set **Xlist** to **L1** and **Ylist** to **L2.** For **Marks** use the first choice, the small square. Press **ZOOM** and press **9** for **ZoomStat** and view the graph.

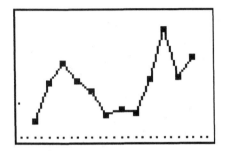

Note: After completing a graph, you should turn the graph **OFF**. Using **2ⁿᵈ** [STAT PLOT], select Plot1 by pressing **ENTER** and highlighting **OFF**. Press **ENTER** and **2ⁿᵈ** [QUIT]. (Note: Turning Plot1 **OFF** is optional. You can leave it ON but leaving it ON will affect other graphing operations.)

Section 2.3

▶ Example 6 (pg. 70) Comparing the Mean, Median, and Mode

Press **STAT** and select **1:Edit**. Clear **L1** and enter the data into **L1**. Press
STAT again and highlight **CALC** to view the Calc Menu.

```
EDIT CALC TESTS
1:1-Var Stats
2:2-Var Stats
3:Med-Med
4:LinReg(ax+b)
5:QuadReg
6:CubicReg
7↓QuartReg
```

Select **1:1-Var Stats** and press **ENTER**. On this line, enter the name of the
column that contains the data. Since you have stored the data in **L1,** simply enter
2ⁿᵈ [L1] ENTER and the first page of the one variable statistics will appear.
(Note: If you did not enter a column name, the default column, which is **L1,**
would be automatically selected.)

```
1-Var Stats L1▮
```

```
1-Var Stats
 x̄=23.75
 Σx=475
 Σx²=13109
 Sx=9.808025715
 σx=9.559680957
↓n=20
```

The first item is the mean, $\bar{x} = 23.75$. Notice the down arrow in the bottom left corner of the screen. This indicates that more information follows this first page. Use the Down Arrow key to scroll through this information. The third item you see on the second page is the median, Med = 21.5.

```
1-Var Stats
↑n=20
 minX=20
 Q₁=20
 Med=21.5
 Q₃=23
 maxX=65
■
```

The TI-83/84 does not calculate the mode but, since this data set is sorted, it is easy to see from the list of data that the mode is 20.

◀

> ▶ Example 7 (pg. 71) Finding a Weighted Mean

Press **STAT** and select **1:Edit**. Clear **L1** and **L2**. Enter the scores into **L1** and the weights into **L2**.

```
L1      L2      L3      2
86      .5      ------
96      .15
82      .2
98      .1
100     .05

L2(6) =
```

Press **STAT** and highlight **CALC** to view the Calc Menu. Select **1:1-Var Stats**, press **ENTER** and press **2ⁿᵈ** **[L1]** **,** **2ⁿᵈ** **[L2]** . Press **ENTER**. (Note: You must place the comma between **L1** and **L2**).

```
1-Var Stats L1,L
2
```

Using **L1** and **L2** in the **1:1-Var Stats** calculation is necessary when calculating a weighted mean. The calculator uses the data in **L1** and the associated weights in **L2** to calculate the average. In this example, the weighted mean is 88.6.

```
1-Var Stats
 x̄=88.6
 Σx=88.6
 Σx²=7885.6
 Sx=
 σx=5.969924623
↓n=1
```

> ▶ Example 8 (pg. 72) Finding the Mean of a Frequency
>   Distribution

Press **STAT** and select **1:Edit**. Clear **L1** and **L2**. Enter the x-values into **L1** and the frequencies into **L2**. Press **STAT**, highlight **CALC**, select **1:1-Var Stats**, press **ENTER**. Next, press **2**nd **[L1]** **⊡** **2**nd **[L2]** **ENTER**.

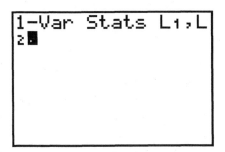

After you press **ENTER**, the sample statistics will appear on the screen. The mean of the frequency distribution described in columns **L1** and **L2** is 41.78. In this example you do not have the actual data. What you have is the frequency distribution of the data summarized into categories. The mean of this frequency distribution is an approximation of the mean of the actual data.

```
1-Var Stats
 x̄=41.78
 Σx=2089
 Σx²=107196.5
 Sx=20.16163259
 σx=19.95899797
↓n=50
```

▶ Exercise 21 (pg. 75) Finding the Mean, Median and Mode

Enter the data into **L1**. Press **STAT** and select **1:1-Var Stats** from the Calc
Menu. Press **2ⁿᵈ L1 ENTER**.

```
1-Var Stats
 x̄=20.6625
 Σx=661.2
 Σx²=14229.92
 Sx=4.280017335
 σx=4.212611274
↓n=32
■
```

The first item in the output screen is the mean, 20.6625. Scroll down through the
output and find the median, Med = 20.05.

```
1-Var Stats
↑n=32
 minX=10.5
 Q₁=18.4
 Med=20.05
 Q₃=23.25
 maxX=30.8
■
```

Although the TI-83/84 does not calculate the Mode, you can use the SORT feature
to order the data. You can then scroll through the data to see if the data set
contains a mode. To sort the data, press **2ⁿᵈ [LIST]** (Note: **List** is found above
the **STAT** key). Move the cursor to highlight **OPS** and select **1:SortA(** and
ENTER.

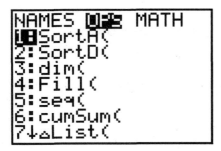

To sort **L1** in ascending order, press **2ⁿᵈ** **[L1]** , close the parentheses and press **ENTER**.

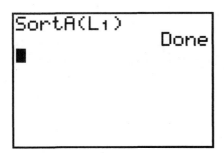

To view the data, press **STAT** and select **1:Edit** and press **ENTER**. Use the down arrow to scroll through **L1** to see if the data has a mode. Look for data values that occur more than once. Record their values and their frequencies. Notice that the value 18.8 occurs twice, 22.1 occurs twice and 26.7 occurs twice. We can conclude that there are three modes: 18.8, 22.1 and 26.7. (Each of these values occurs twice. All the other values occur only once.)

▶ Exercise 49 (pg. 78) Finding the Mean of Grouped Data

Press **STAT** and select **1:Edit**. Clear **L1** and **L2**. Enter the midpoints of each
Age group into **L1** and the frequencies into **L2**.

L1	L2	L3 3
4.5	55	
14.5	70	
24.5	35	
34.5	56	
44.5	74	
54.5	42	
64.5	38	

L3(1)=

Press **STAT** and highlight **CALC** to view the Calc Menu. Select **1:1-Var Stats**,
press **ENTER** and type in **2ⁿᵈ L1** , **2ⁿᵈ L2**. The mean of this grouped data is
35.76.

```
1-Var Stats
 x̄=35.75944584
 Σx=14196.5
 Σx²=698629.25
 Sx=21.96014969
 σx=21.93247463
↓n=397
■
```

▶ Exercise 53 (pg. 79) Construct a Frequency Histogram

To prepare to construct the histogram, press the **Y=** key and clear all the Y-registers. Press **STAT** and select **1:Edit**. Clear **L1** and enter the data into **L1**. Press **2nd** [STAT PLOT] and select **1:Plot 1** and press **ENTER**. Turn **ON** Plot 1 and set the **Type** to **Histogram** and press **ENTER**. Move the cursor to **Xlist** and set this to **L1**. Move the cursor to **Freq** and set it equal to **1.** (Note: The cursor may be in **ALPHA** mode with a flashing **A** . Press **ALPHA** to return to the solid rectangular cursor and type in **1**.)

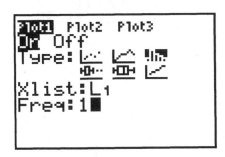

Press **ZOOM** and **9** to select **ZoomStat**. The histogram that is displayed has 7 classes. To change to 5 classes, press **Window**. Notice that **Xmin** = 62 and **Xmax** = 78.33. Change **Xmax** to 79, which is the next whole number greater than 78.33. To determine a value for **Xscl**, you must calculate (**Xmax - Xmin**)/5. Press **2nd** [QUIT] to close out the Window Menu. (Note: QUIT is found above the **MODE** key). Calculate (79 - 62)/5 and round the resulting value, 3.4, to 3. Press **Window** and set **Xscl** = 3. Press **GRAPH** and view the histogram with 5 classes. Notice that one of the bars is too tall to fit completely on the screen. Press **Window** and set **Ymax** = 9, which is the next whole number greater than 8.19. Press **GRAPH** again. As you can see from the graph, the histogram appears to be symmetric.

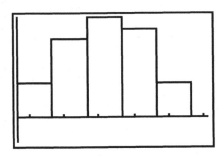

Exercise 60 (pg. 80) Data Analysis

Press **STAT** and select **1:Edit**. Clear **L1** and then enter the data. Press **STAT** and highlight **CALC**. Select **1:1-Var Stats** and press **2nd L1 ENTER** to obtain values for the mean and median.

Since the TI-83/84 does not do a stem-and-leaf plot, you can use a histogram to get a picture of the data. To prepare to construct the histogram, press the **Y=** key and clear all the Y-registers. Press **2nd [STAT PLOT]** , select **1:Plot 1** and press **ENTER**. Turn ON **Plot 1**. Set **Type** to **Histogram**. The **Xlist** is **L1** and the **Freq** is **1**. Press **ZOOM** and scroll down to **9:ZoomStat** and press **ENTER** and a histogram will appear on the screen.

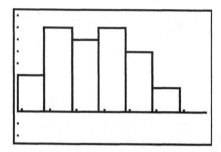

To construct a histogram that would be similar to a stem-and-leaf plot that uses one row per stem, press **WINDOW** and set **Xmin = 10**, **Xmax = 100** and **Xscl = 10**. Press **GRAPH** to display the new histogram.

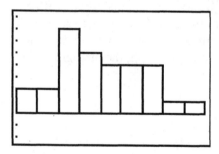

You can press **TRACE** and scroll through the bars of the histogram. Min and Max values for each bar will appear along with the number of data points in each class. You can use this information to determine the location of the mean and the median.

There appears to be some positive skewness in the data. (Note: This is more easily seen in the second of the two histograms.)

Section 2.4

▸ Example 5 (pg. 86) Finding the Standard Deviation

Press **STAT** and select **1:Edit**. Clear **L1** and enter the data into **L1**. Press **STAT** and highlight **CALC** to display the Calc Menu. Select **1: 1-Var Stats** and press **2ⁿᵈ L1 ENTER**. The sample standard deviation is Sx, 5.089373342

```
1-Var Stats
 x̄=33.72916667
 Σx=809.5
 Σx²=27899.5
 Sx=5.089373342
 σx=4.982216639
↓n=24
```

◀

▸ Example 9 (pg. 90) Standard Deviation of Grouped Data

Press **STAT** and select **1:Edit**. Clear **L1** and **L2**. Enter the x-values into **L1** and the frequencies into **L2**. Press **STAT** and highlight **CALC** to select the Calc Menu. Select **1:1-Var Stats** and press **ENTER**. Type in **2ⁿᵈ L1 , 2ⁿᵈ L2** and press **ENTER**. From the statistics displayed, the mean is 1.82 and the sample standard deviation is 1.722.

```
1-Var Stats
 x̄=1.82
 Σx=91
 Σx²=311
 Sx=1.722480414
 σx=1.705168613
↓n=50
```

◀

▶ Exercise 19 (pg. 94) Comparing Two Datasets

Press **STAT** and select **1:Edit**. Clear **L1** and **L2**. Enter the Los Angeles Data into **L1** and the Long Beach Data into **L2**. Press **STAT** and highlight **CALC**. Select **1:1-Var Stats** and press **2nd L1 ENTER**. The sample standard deviation for the Los Angeles data is Sx = 6.111.

```
1-Var Stats
 x̄=26.25555556
 Σx=236.3
 Σx²=6502.97
 Sx=6.111282826
 σx=5.761772704
↓n=9
■
```

To find the range, scroll down through the display and find **minX = 18.3.** Continue scrolling through the display and find **maxX = 35.9.** To calculate the range, simply type in **35.9 - 18.3** and press **ENTER**. The range = 17.6.

```
minX=18.3
Q₁=20.55
Med=26.1
Q₃=31.85
maxX=35.9
35.9-18.3
              17.6
```

To find the variance, you must square the standard deviation. Type in 6.111 and press the x^2 key. The variance is 37.344.

```
Med=26.1
Q₃=31.85
maxX=35.9
35.9-18.3
             17.6
6.111²
       37.344321
```

For the Long Beach Data, press **STAT** , highlight **CALC**, select **1:1-Var Stats**, press **ENTER** and press 2nd **L2.**

```
1-Var Stats L₂█
```

Press **ENTER**. The sample standard deviation for the Long Beach Data is Sx = 2.952.

```
1-Var Stats
 x̄=22.87777778
 Σx=205.9
 Σx²=4780.25
 Sx=2.952023788
 σx=2.783194718
↓n=9
█
```

Use the same procedure as you used for the Los Angeles Data, to find the range (8.7) and the variance (8.71) for the Long Beach Data.

The annual salaries in Los Angeles are more variable than the salaries in Long Beach.

◀

Section 2.5

▶ **Example 2 (pg. 103)** Finding Quartiles

Press **STAT** and select **1:Edit**. Clear **L1** and enter the data into **L1**. Press **STAT** and highlight **CALC**. Select **1:1-Var Stats** and press **2nd L1 ENTER**. Scroll down through the descriptive statistics. You will see the first quartile: **Q1** = **21.5**, the second quartile (the median): **Med = 23** and the third quartile: **Q3 = 28.**

```
1-Var Stats
↑n=25
 minX=15
 Q₁=21.5
 Med=23
 Q₃=28
 maxX=30
```

◀

▶ Example 4 (pg. 105) Drawing a Box-and-Whisker-Plot

To prepare to construct the Box-and-Whisker plot, press the **Y=** key and clear all the Y-registers. Press **STAT** and select **1:Edit**. Clear **L1** and enter the data from Example 1 on pg. 102 in your textbook. Press **2ⁿᵈ** [STAT PLOT]. Select **1:Plot 1** and press **ENTER**. Turn On **Plot 1**. Using the right arrow (you can not use the down arrow to drop to the second line), scroll through the **Type** options and choose the second boxplot which is the middle entry in row 2 of the **TYPE** options. Press **ENTER**. Move to **Xlist** and type in **2ⁿᵈ L1**. Press **ENTER** and move to **Freq**. Set **Freq** to **1**. If **Freq** is set on **L2**, press **ALPHA** to return the cursor to a flashing solid rectangle and type in **1**. Press **ZOOM** and **9** to select **ZoomStat**. The Boxplot will appear on your screen.

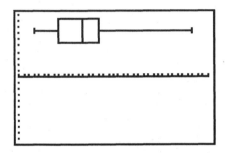

If you press **TRACE** and use the left and right arrow keys, you can display the five values that represent the five-number summary of the data.

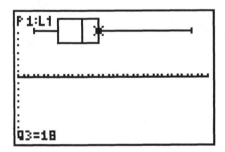

Notice in the above screen display that the trace cursor is on the right side of the box, which represents the third quartile. Also, at the bottom of the screen it is noted that Q3=18.

◀

▶ Exercise 1 (pg. 109) Quartiles and a Box-and-Whisker-Plot

To prepare to construct the Box-and-Whisker plot, press the **Y=** key and clear all the Y-registers. Press **STAT** and select **1:Edit**. Clear **L1** and enter the data . Press **2ⁿᵈ** [STAT PLOT]. Select **1:Plot 1** and press **ENTER**. Turn On **Plot 1**. Using the right arrow, scroll through the **Type** options and choose the second boxplot which is the second entry in row 2 of the **TYPE** options. Press **ENTER**. Move to **Xlist** and type in **2ⁿᵈ** **L1**. Press **ENTER** and move to **Freq**. Set **Freq** to **1**. If **Freq** is set on **L2**, press **ALPHA** to return the cursor to a flashing solid rectangle and type in **1**. Press **ZOOM** and **9** to select **ZoomStat.** The Boxplot will appear on your screen. If you press **TRACE** and use the right and left arrows, you can display **Min, Q1, Med, Q3** and **Max** . Notice **Med**=6 for this example.

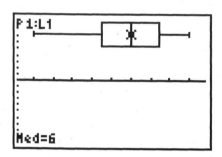

> **Exercise 23 (pg. 110)** Quartiles and a Box-and-Whisker-Plot

To prepare to construct the Box-and-Whisker plot, press the **Y=** key and clear all the Y-registers. Press **STAT** and select **1:Edit**. Clear **L1** and enter the data. Press **2ⁿᵈ** [STAT PLOT]. Select **1:Plot 1** and press **ENTER**. Turn On **Plot 1**. Using the right arrow, scroll through the **Type** options and choose the second boxplot which is the second entry in row 2 of the **TYPE** options. Press **ENTER**. Move to **Xlist** and type in **2ⁿᵈ L1**. Press **ENTER** and move to **Freq**. Set **Freq** to **1**. If **Freq** is set on **L2**, press **ALPHA** to return the cursor to a flashing solid rectangle and type in **1**. Press **ZOOM** and **9** to select **ZoomStat**. The Boxplot will appear on your screen. If you press **TRACE** and use the right and left arrows, you can display **Min, Q1, Med, Q3** and **Max**.

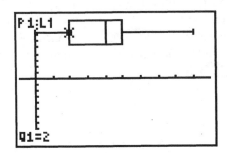

▶ Technology (pg. 123) Monthly Milk Production

Exercises 1-2: Press **STAT** and select **1:Edit**. Clear **L1** and enter the data into **L1**. Press **STAT** and highlight **CALC**. Select **1:1-Var Stats** and press 2nd **L1** **ENTER**. The sample mean and sample standard deviation can be found in the output display.

Exercises 3-4. You will use the histogram of the data to construct the frequency distribution. To prepare to construct the histogram, press the **Y=** key and clear all the Y-registers. To construct a histogram, press 2nd [STAT PLOT]. Select **1: Plot 1** and **ENTER**. Turn ON **Plot 1** by highlighting **On** and pressing **ENTER**. Scroll through the **Type** icons, highlight the **Histogram** and press **ENTER**. Move to **Xlist** and type in 2nd **L1**. Move to the **Freq** entry and type in **1.** (Note: If the Freq entry is **L2**, Press **ALPHA** and enter **1**). To view a histogram of the data, press **ZOOM** and **9**. To adjust the histogram, press **Window**. Set **Xmin = 1000** and set **Xscl = 500.** (All other entries in the Window Menu can remain unchanged). Press **GRAPH** and then press **TRACE**.

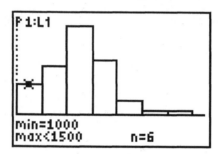

The minimum and maximum values of the first category are displayed. (**1000 and < 1500**). The midpoint of the first category is (1000+1499)/2, which is **1249.5.** The frequency for the first category is **n = 6.** Trace through the histogram and set up a frequency distribution for the data:

Category	Midpoint	Frequency
1000 – 1499	124 9.5	6

Exercise 5. Using the values that you found in Exercises 1 and 2, calculate the lower and upper endpoints of the one and two standard deviation intervals: $(\bar{x} \pm 1s)$ and $(\bar{x} \pm 2s)$. Next, press **2ⁿᵈ** [LIST]. Highlight **OPS** and select **1:SortA(** , press **ENTER**. Press **2ⁿᵈ** **[L1]** and close the parentheses. Press **ENTER**. Press **STAT**, select **1:Edit** and press **ENTER**. Scroll through the data in **L1** and count the number of data points that fall within the one and two standard deviation intervals: $(\bar{x} \pm 1s)$ and $(\bar{x} \pm 2s)$. To find the percentage of data points that lie in each of these intervals, calculate: (number of data points in each interval/ 50) * 100. Compare these percentages with the percentages stated in the Empirical rule.

Exercises 6 - 7. Using the frequency distribution you created, enter the midpoints into **L2** and the frequencies into **L3**. Press **STAT** , highlight **CALC**. Select **1:1-Var Stats** and press **ENTER**. Type in **2ⁿᵈ L2** , **2ⁿᵈ L3** and press **ENTER**. The mean and standard deviation for the frequency distribution will be displayed on the screen.

Exercise 8: Compare the actual values for the mean and standard deviation that you found in Exercises 1 and 2 to the estimates for the mean and standard deviation that you found in Exercises 6 and 7.

CHAPTER

Probability

3

Section 3.1

▶ **Law of Large Numbers (pg. 138)**

You can use the TI-83/84 to simulate tossing a coin 150 times.(Note: The scatterplot on pg.138 displays the results of simulating a coin toss 150 times.) In this simulation, we will designate "0" as Heads and "1" as Tails.

Press **MATH**, highlight **PRB**, and select **5:randInt(** and press **ENTER**. The **randInt(** command requires a minimum value, (which is 0 for this simulation), a maximum value (which is 1), and the number of trials (150). In the **randInt(** command type in **0** ⬜ **1** ⬜ **150.**

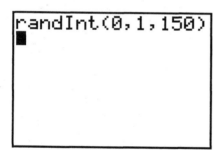

```
randInt(0,1,150)
■
```

Press **ENTER**. It will take a few seconds for the calculator to generate 150 tosses. Notice, in the upper right hand corner a flashing ⬜, indicating that the calculator is working. When the simulation has been completed, a string of **0's** and **1's** will appear on the screen followed by **….**, indicating that there are more numbers in the string that are not shown.

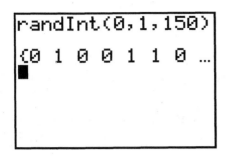

```
randInt(0,1,150)
{0 1 0 0 1 1 0 ...
■
```

Press **STO** and 2nd **[L1]** **ENTER**. This will store the string of numbers in **L1**. Press 2nd **[LIST]** and highlight **MATH**. Select **sum(** and type in **L1** and close the parentheses.

```
randInt(0,1,150)
{0 1 0 0 1 1 0 ...
Ans→L₁
{0 1 0 0 1 1 0 ...
sum(L₁)
                71
```

The sum of **L1** equals the number of Tails in the list. For this particular simulation, the sum is 71. The number of Heads is equal to (150 - no. of Tails). The proportion of Heads is (no. of Heads / 150). How close is this proportion to 50% ?

◀

Section 3.2

▶ Exercise 36d (pg.159) Birthday Problem

To simulate this problem, press **MATH**, highlight **PRB**, select **5:randInt(** and enter 1 ⌐, 365 ⌐, 24.

```
randInt(1,365,24
)
{250 364 159 26…
```

Store the data in **L1** by pressing **STO** and 2nd **[L1]** **ENTER**. To see if there are at least two people with the same birthday, you must look for a matching pair of numbers in **L1**. To do this, press 2nd **[LIST]**, highlight **OPS** and select **1:SortA(** and press **ENTER**. The column you want to sort in ascending order is **L1**, so type 2nd **[L1]** into the sort command and press **ENTER**.

```
randInt(1,365,24
)
{292 351 153 60…
Ans→L1
{292 351 153 60…
SortA(L1)
            Done
```

Press **STAT** and select **1:EDIT**, press **ENTER** and scroll down through **L1** and check for matching numbers. If you find any matching numbers, that means that at least 2 people in your simulation have the same birthday.

```
L1      L2      L3      1
[15]    ------  ------
15
23
44
60
64
74
L1(1)=15
```

Notice in this simulation, 15 is listed twice. This represents two people with the same birthday, January 15 (the 15th day of a year). That means that for your first simulation you have found at least two people with the same birthday. Since you have found this matching pair right at the beginning of **L1**, you do not need to look further into the data in **L1**. Repeat this simulation process nine more times, each time checking to see if you have a matching pair.

How many of your simulations resulted in at least one matching pair? Suppose you found a matching pair in 5 out of your 10 simulations. That means that your estimate of the probability of finding at least 2 people with the same birthday in a room of 24 people is "5 out of 10" or 50%.

Section 3.4

▸ Example 1 (pg. 172) Finding the Number of Permutations

How many different ways can the first row of the Sudoku grid be filled? To find
the number of different arrangements of the numbers 1 through 9, you must
calculate the number of permutations of **9** numbers taken **9** at a time. The
formula **nPr** is used with **n = 9** (the number of digits) and **r = 9** (the number of
digits you will be selecting). So, the formula is **9P9**.

Press **9**, **MATH**, highlight **PRB** and select **2:nPr** and **ENTER**.

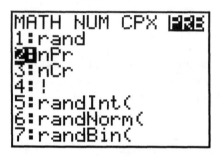

Now press **9** and **ENTER**. The answer, 362880, appears on the screen.

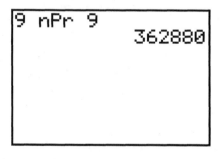

▶ Example 3 (pg. 173) More Permutations

How many ways can forty-three race cars finish first, second and third in a race? Use the permutation formula with **n = 43** and **r = 3**. Enter **43**, press **MATH**, highlight **PRB**, select **2:nPr** and press **ENTER**. Now enter **3** and press **ENTER**.

▶ Example 4 (pg. 174) Distinguishable Permutations

How many distinguishable ways can the 6 one-story houses, 4 two-story houses and 2 split-level houses be arranged? To calculate $\dfrac{12!}{6!4!2!}$ you will use the factorial function **(!).** Enter the first value, **12**, press **MATH**, highlight **PRB** and select **4:!**. Then press **÷** . Open the parentheses by pressing **(**. Enter the next value, **6**, press **MATH**, **PRB**, and select **4:!** . To multiply by 4!, press **x** and enter the next value, **4**. Press **MATH**, **PRB**, and select **4:!**. To multiply by 2!, press **x** and enter the next value, **2**. Press **MATH**, **PRB**, and select **4:!**. Close the parentheses **)** and press **ENTER**.

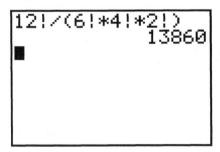

▶ Example 5 (pg. 175) Finding the Number of Combinations

To calculate the number of different combinations of four companies that can be selected from 16 bidding companies you will use the combination formula: **nCr,** where n = the total number of items in the group and r = the number of items being selected. In this example, the formula is **16C4**. Enter the first value, **16,** press **MATH**, highlight **PRB** and select **3: nCr,** and enter the second value, **4.** Press **ENTER** and the answer will be displayed on the screen..

```
16 nCr 4
                1820
```

▶ Example 8 (pg. 177) Finding Probabilities

To calculate the probability of being dealt five diamonds from a standard deck of 52 playing cards, you must calculate: $\dfrac{13C_5}{52C_5}$

To calculate the numerator, enter the first value, **13**, press **MATH**, highlight **PRB** and select **3:nCr** and enter the next value, **5**. Next press **÷** and enter the first value in the denominator, **52**, press **MATH**, highlight **PRB** and select **3:nCr**, enter the next value, **5**. Press **ENTER** and the answer will be displayed on your screen.

```
13 nCr 5/52 nCr
5
    4.951980792E-4
```

Notice that the answer is written in scientific notation. To convert to standard notation, move the decimal point 4 places to the left: 0.000495.

▸ Exercise 50 (pg. 181) Probability

a. To calculate $_{200}C_{15}$, enter the first value, **200**, press **MATH**, highlight **PRB** and select **3:nCr**, enter the next value, **15** and press **ENTER**.

b. Calculate $_{144}C_{15}$ (follow the steps for part a.).

c. The probability that *no* minorities are selected is equal to the probability that the committee is composed completely of non-minorities. This probability can be calculated with the following formula: $\dfrac{_{144}C_{15}}{_{200}C_{15}}$

(Use the answer from part b for the numerator and the answer from part a for the denominator.)

◂

▶ Technology (pg. 191) Composing Mozart Variations

Exercise 1: The player has 11 phrases to choose from for the 14 bars and 2 phrases for the remaining 2 bars. The total number of phrases to choose from is:

Exercise 2: Us the Fundamental Principle of Counting to calculate the number of possible variations. For 14 of the bars there are 11 choices resulting in 11^{14} choices. For the remaining 2 bars, there are 2 choices resulting in 2^2 choices. For the total number of choices, multiply 11^{14} by 2^2.

Exercise 3: To select one number from 1 to 11, press **MATH**, highlight **PRB** and select **5:randInt(** and enter **1** **,** **11** **)** and press **ENTER**. One random number between 1 and 11 will be displayed on the screen.

```
randInt(1,11)
              5
```

Exercise 3a: Each number from 1 to 11 has an equal chance of being selected. So each of the 11 possibilities has equal probability. What is this theoretical probability for each number from 1 to 11? (Use this to answer the question in Exercise 4.)

Exercise 3b: To select 100 integers between 1 and 11, press **MATH**, highlight **PRB** and select **5:randInt(** and enter **1** **,** **11** **,** **100** **)** and press **ENTER**. Store the results in **L1** by pressing **STO**, **2ⁿᵈ** **[L1]**, **ENTER**. Next you can create a histogram and use it to tally your results. To create the histogram, press **2ⁿᵈ** **[STAT PLOT]**, select **1:Plot 1** and press **ENTER**. Turn **ON** Plot 1, set **Type** to **Histogram.** Set **Xlist** to **L1** and **Freq** to **1.** Press **ZOOM** and select **9** for **ZoomStat.** Press **WINDOW** and set **Xscl = 1,** then press **GRAPH**. Use **TRACE** to scroll through the bars of the histogram.

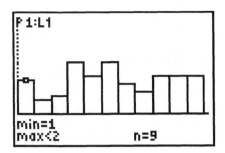

The first bar represents the number of times that a "**1**" occurred in the 100 tosses. (Notice on the screen: min = 1and n=9). Use the right arrow to scroll through the bars in the histogram. As the cursor moves from one bar in the histogram to the next bar, the "min =" will indicate the current x-value and the "n=" will indicate the corresponding frequency. Record the x-values and the corresponding frequencies in a table. For example, using the histogram in the above diagram, the frequency table would begin with X = 1, n = 9. Convert each frequency into a relative frequency by dividing each frequency by 100, and use the relative frequencies to answer the question in Exercise 4. Compare the relative frequencies for each number from 1 to 11 with the theoretical probabilities you obtained in part 3a.

Exercise 4a: The theoretical probability of selecting phrase 6, 7 or 8 for the 1st bar is the sum of the theoretical probabilities you found in part 3a. Raise this answer to the 14th power to find the theoretical probability of selecting phrase 6, 7 or 8 for all 14 bars.

Exercise 4b: Use the relative frequencies found in part 3b. for phrases 6, 7 or 8 to estimate the probability of selecting phrases 6, 7 or 8 for the 1st bar. Raise this answer to the 14th power to estimate the probability of selecting phrase 6, 7 or 8 for all 14 bars.

Exercise 5: To select 2 random numbers from 1 to 6, press **MATH**, highlight **PRB** and select **5:randInt(** and enter 1 **,** 6 **,** 2) and press **ENTER**. Add the two numbers in the display and subtract **1** from the total.

Exercise 5a: First list all possible pairs. (There are 36 pairs: (1,1), (1,2), (1,3)....). For each pair, find the sum and subtract "1" to find a total. Make a frequency table listing all the possible totals from 1 to 11. Next to each total record the number of times that total occurred. Lastly, convert each of these frequencies to relative frequencies by dividing by 36. These relative frequencies are the theoretical probabilities. Use them to answer the question in Exercise 6a.

Exercise 5b: To select 100 totals, create 2 columns of 100 random numbers. Press **MATH**, highlight **PRB** and select **5:randInt(** and enter **1** , **6** , **100**) and press **ENTER**. Press **STO** **2nd** **[L1]** and **ENTER** to store this first string of numbers in **L1**. Next create a second set of 100 numbers using the same procedure. Press **STO** **2nd** **[L2]** and **ENTER** to store this second string of numbers in **L2**. Now you can calculate the sum of each row and subtract **1**. Press **STAT** and select **1:EDIT**. Move the cursor to highlight **L3**.

```
L1        L2        L3       3
1         4         ------
3         2
6         4
2         3
1         1
1         2
3         6

L3 =
```

Press **ENTER** and the cursor will move to the bottom of the screen. On the bottom line next to **L3 =**, type in 2nd **[L1]** + 2nd **[L2]** - 1 and press **ENTER**. The results should appear in **L3**.

```
L1        L2        L3       3
1         4         4
3         2         4
6         4         9
2         3         4
1         1         1
1         2         2
3         6         8

L3(1)=4
```

Create a histogram of **L3** and use it to tally your results (follow the steps in Exercise 3 to create the histogram). Use **TRACE** to find the frequencies for each X-value and record your results in a frequency table. Convert each frequency to a relative frequency and use the relative frequencies to answer the question in Exercise 6b.

Discrete Probability Distributions

Section 4.1

▶ Example 5 (pg. 198) Mean of a Probability Distribution

Press **STAT** and select **1:EDIT**. Clear **L1** and **L2**. Enter the X-values into **L1** and the P(x) values into **L2**. Press **STAT** and highlight **CALC**. Select **1:1-Var Stats,** press **ENTER** and press 2nd **[L1]** **,** 2nd **[L2]** **ENTER** to see the descriptive statistics. The mean score is 2.94.

```
1-Var Stats
 x̄=2.94
 Σx=2.94
 Σx²=10.26
 Sx=
 σx=1.271377206
↓n=1
```

▶ Example 6 (pg. 199) The Variance and Standard Deviation

This example is a continuation of Example 5. The data is displayed in the table on pg. 199. Enter the X-values into **L1** and the P(x) values into **L2**. Press **STAT** highlight **CALC**. Select **1:1-Var Stats,** press **ENTER** and press 2nd **[L1]** **,** 2nd **[L2]** **ENTER**. The population standard deviation , σx, is 1.27.

```
1-Var Stats
 x̄=2.94
 Σx=2.94
 Σx²=10.26
 Sx=
 σx=1.271377206
↓n=1
■
```

To find the variance, you must use the value of the standard deviation. Since the variance is equal to the standard deviation squared, type in **1.2714** and press the x^2 key. The population variance is **1.616**.

```
 Σx=2.94
 Σx²=10.26
 Sx=
 σx=1.271377206
↓n=1
1.2714²
          1.61645796
```

Section 4.2

▶ Example 4 (pg. 210)   Binomial Probabilities

To find a binomial probability you will use the binomial probability density
function, **binompdf(n,p,x).** For this example, n = 100, p = .59 and x = 65. Press
2ⁿᵈ [DISTR]. Scroll down through the menu to select **binompdf(** and press
ENTER . Type in **100** |, **.59** |, **65**) and press **ENTER**. The answer, **0.0391**,
will appear on the screen.

```
binompdf(100,.59
,65)
        .0391071795
-
```

▶ Example 7 (pg. 213) Graphing Binomial Distributions

Construct a probability distribution for a binomial probability model with n = 6 and p = .59. Press **STAT**, select **1:EDIT** and clear **L1** and **L2**. Enter the values 0 through 6 into **L1**. Press **2nd [QUIT]**. To calculate the probabilities for each X-value in **L1**, first change the display mode so that the probabilities displayed will be rounded to 3 decimal places. Press **MODE** and change from **FLOAT** to **3** and press **2nd [QUIT]**. Next press **2nd [DISTR]** and select **binompdf(** and type in **6 , .59)** and press **ENTER**. Store these probabilities in **L2** by pressing **STO 2nd [L2] ENTER**.

```
binompdf(6,.59)
{.005 .041 .148…
Ans→L₂
{.005 .041 .148…
```

To prepare to construct the probability histogram, press the **Y=** key and clear all the Y-registers. To graph the binomial distribution, press **2nd [STAT PLOT]** and press **ENTER**. Turn **ON** Plot 1, select **Histogram** for **Type**, set **Xlist** to **L1** and set **Freq** to **L2**. Press **ZOOM** and select **9** for **ZoomStat**. Adjust the graph by pressing **WINDOW** and setting **Xmin = 0, Xmax = 7, Xscl = 1, Ymin = 0** and **Ymax = .32.** Choosing an Xmax=7, leaves some space at the right of the graph in order to complete the histogram. The Ymax value was selected by looking through the values in **L2** and then rounding the largest value UP to a convenient number. Press **GRAPH** to view the histogram.

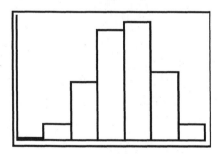

▶ Exercise 15 (pg. 216) Binomial Probabilities

A multiple-choice quiz consists of five questions and each question has four possible answers. If you randomly guess the answer to each question, the probability of getting the correct answer is 1 in 4, so p = .25 and n = 5.

a.   To calculate P(x = 3), press **2ⁿᵈ [DISTR]** and select **binompdf(** and type in **5 [,] .25 [,] 3 [)]** and press **ENTER**.

b-c. To calculate inequalities, such as P(X ≥ 3) or P(X < 3) , you can use the cumulative probability command: **binomcdf (n,p,x)**. This command accumulates probability starting at X = 0 and ending at a specified X-value. To calculate the probability of guessing at least three answers correctly, that is P(X ≥ 3), you must find P(X ≤ 2) and subtract from 1. Press **2ⁿᵈ [DISTR]** and select **binomcdf(** by scrolling through the options and selecting **binomcdf(** . Type in **5 [,] .25 [,] 2 [)]** and press **ENTER**. The result, P(X ≤ 2) = .896.

```
binomcdf(5,.25,2
)
          .896484375
1-.896
               .104
```

 This result is the accumulated probability that X = 0,1 or 2, since this command, **binomcdf**, accumulates probability starting at X = 0. You want P(X ≥ 3) and this is the *complement* of P(X ≤ 2). So P(X ≥ 3) = 1 - P(X ≤ 2) which is .104. The probability of guessing *less than* 3 answers correctly is P(X ≤ 2) which equals 0.896.

◀

Section 4.3

> **Example 1 (pg. 222)** The Geometric Distribution

Suppose the probability that you will make a sale on any given telephone call is 0.23. To find the probability that your *first* sale on any given day will occur on your fourth or fifth telephone call of the day, press **2ⁿᵈ** **[DISTR]** and select **geometpdf(** and type in **.23** | **,** | **2ⁿᵈ** | **{** | **4** | **,** | **5** | **2ⁿᵈ** | **}** | **)** . This command will display 2 probabilites: P(X = 4) and P(X = 5). The probability that your first sale of the day will occur on your fourth or fifth sales call of the day is .105003 + .080852.

```
geometpdf(.23,{4
,5})
{.10500259 .080…
■
```

> **Example 2 (pg. 223)** The Poisson Distribution

Use the command **poissonpdf (μ,x)** with $\mu = 3$ and $x = 4$. Press **2ⁿᵈ** **[DISTR]** and select **poissonpdf(** and type in **3** | **,** | **4** | **)** and press ENTER. The answer will appear on the screen.

```
poissonpdf(3,4)
         .1680313557
```

▶ Exercise 21 (pg. 227) Poisson Distribution

This is a Poisson probability problem with $\mu = 3$.

a.   To find the probability that exactly 5 businesses will file bankruptcy in any given minute,   press **2ⁿᵈ** **[DISTR]** and select **poissonpdf(** . Type in **3** **[,]** **5** **[)]** and press **ENTER**.

```
poissonpdf(3,5)
        .1008188134
```

b.   Use the cumulative probability command **poissoncdf(**. This command accumulates probability starting at $X = 0$ and ending at the specified X-value. To calculate the probability that at least 5 businesses will fail in any given minute, that is $P(X \geq 5)$, you must find $P(X \leq 4)$ and subtract from 1. Press **2ⁿᵈ** **[DISTR]** , select   **poissoncdf(** and type in **3** **[,]** **4** **[)]** and press **ENTER**. The result, $P(X \leq 4)$, is 0.815. This result is the accumulated probability that $X = 0,1,2, 3$ or 4. $P(X \geq 5)$ is the *complement* of $P(X \leq 4)$. So, $P(X \geq 5)$ is $1 - 0.815$ or 0.185.

```
poissoncdf(3,4)
        .8152632446
1-.815
             .185
```

c.   To find the probability that more than 5 businesses will fail in any given hour, find P(X > 5). So, first calculate the *complement*, P(X ≤ 5), and then subtract the answer from 1.

```
Poissoncdf(3,5)
           .9160820581
1-.916
                 .084
■
```

P(X > 5) = 1 – 0.916 = 0.084.

> Technology (pg. 237)

Exercise 1: Create a Poisson probability distribution with $\mu = 4$ for $X = 0$ to 20. Press **STAT** and select **1:EDIT** . Clear **L1** and **L2**. Enter the integers 0,1,2,3,…20 into **L1** and press **2ⁿᵈ [QUIT]**.

For this example, it is helpful to display the probabilities with 3 decimal places. Press **MODE**. Move the cursor to the 2ⁿᵈ line and select **3** and press **ENTER**.

Press **2ⁿᵈ [DISTR]** and select **B:poissonpdf(** . Type in **4** ⌷ **2ⁿᵈ [L1]** and press **ENTER** .

```
poissonpdf(4,L₁)
{.018 .073 .147…
■
```

A partial display of the probabilities will appear. Press **STO** and **2ⁿᵈ [L2] ENTER**. Press **STAT**, and select **1:EDIT.** **L1** and **L2** will be displayed on the screen.

L1	L2	L3	1
0.000	.018	------	
1.000	.073		
2.000	.147		
3.000	.195		
4.000	.195		
5.000	.156		
6.000	.104		

L1(1)=0

Notice that the X-values in **L1** now have 3 decimal places. This is a result of setting the **MODE** to **3.** Each value in **L2** is the Poisson probability associated with the X-value in **L1**. So, for example, $P(X = 2)$ is .147. Scroll through **L2** and compare the probabilities in **L2** with the heights if the corresponding bars in the frequency histogram on page 237.

Exercises 3 - 5: Use another technology tool to generate the random numbers. The TI-83 cannot generate Poisson random data.

Exercise 6: For this exercise, $\mu = 5$ and $X = 10$. Press **2nd** **[DISTR]** and select **B:poissonpdf(** . Type in **5** , **10** and press ENTER .

Exercise 7: Use the probability distribution that you generated in Exercise 1.

a. Sum the probabilities in **L2** that correspond to X- values of 3, 4 and 5 in **L1**.

b. Sum the probabilities in **L2** that correspond to X- values of 0,1,2,3 and 4 in **L1**. This is $P(X \leq 4)$. Subtract this sum from **1** to get $P(X > 4)$.

c. Assuming that the number of arrivals per minute are independent events, raise $P(X > 4)$ to the fourth power.

Normal Probability Distributions

CHAPTER

5

Section 5.2

▶ Example 3 (pg. 255) Normal Probabilities

Suppose that cholesterol levels of American men are normally distributed with a mean of 215 and a standard deviation of 25. If you randomly select one American male, calculate the probability that his cholesterol is less than 175, that is P(X < 175).

The TI-83/84 has two methods for calculating this probability.

Method 1: **Normalcdf(***lowerbound, upperbound,* μ *,* σ **)** computes probability between a *lowerbound* and an *upperbound.* In this example, you are computing the probability to the left of 175, so 175 is the *upperbound.* In examples like this, where there is no *lowerbound,* you can always use *negative infinity* as the *lowerbound.* Negative infinity is specified by (-) 1 2nd [EE] 9 9 (Note: **EE** is found above the comma ,). Try entering   −1 EE 99 into your calculator.

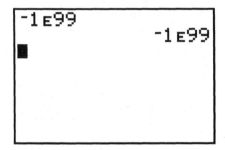

Now, to calculate P(X < 175), press **2nd** **[DISTR]** and select **2:normalcdf(** and type in **-1E99** , 175 , 215 , 25 **)** and press **ENTER**.

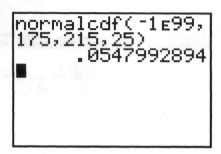

Technical note: Theoretically, the normal probability distribution (the bell-shaped curve) extends infinitely to the right and left of the mean (see Textbook pg. 240). In this particular problem, P(X < 175), you do not necessarily have to use negative infinity (-1 EE 99) as your *lowerbound*. If you look at this example in your textbook on pg. 255, the *lowerbound* is set at 0. This is a perfectly fine selection since no individual will have a cholesterol level less than 0.

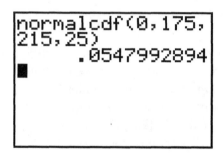

Notice that your results are the same in both of the above screens.

Method 2: This method calculates P(X < 175) and also displays a graph of the probability distribution. You must first clear the Y-registers and turn OFF all STATPLOTS. Next, set up the WINDOW so that the graph will be displayed properly. You will need to set Xmin equal to (μ - 3 σ) and Xmax equal to (μ + 3 σ). Press **WINDOW** and set **Xmin** equal to (μ - 3 σ) by entering **215 - 3 * 25.** Press **ENTER** and set **Xmax** equal to (μ + 3 σ) by entering **215 + 3 * 25.** Set **Xscl** equal to σ.

Setting the Y-range is a little more difficult to do. A good "rule - of - thumb" is to set **Ymax** equal to .5 / σ. For this example, type in **.5 / 25** for **Ymax.**

```
WINDOW
 Xmin=140
 Xmax=290
 Xscl=25
 Ymin=-.005
 Ymax=.5/25█
 Yscl=1
 Xres=1
```

Use the Up Arrow key to highlight **Ymin**. A good value for **Ymin** is **(-) Ymax /
4** so type in ☐**(-)** **.02** **/** **4**.

```
WINDOW
 Xmin=140
 Xmax=290
 Xscl=25
 Ymin=-.02/4█
 Ymax=.02
 Yscl=1
 Xres=1
```

Press **2ⁿᵈ** **[QUIT]**. Clear all the previous drawings by pressing **2ⁿᵈ** **[DRAW]**
and selecting **1:ClrDraw** and pressing **ENTER** **ENTER**. Now you can draw
the probability distribution. Press **2ⁿᵈ** **[DISTR]**. Highlight **DRAW** and select
1:ShadeNorm(and type in **-1E99** ☐**,** **175** ☐**,** **215** ☐**,** **25** ☐**)** and press **ENTER**.
The output displays a normal curve with the appropriate area shaded in and its
value computed.

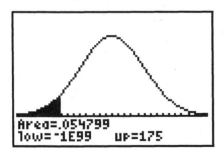

▶ Exercise 13 (pg. 257) Finding Probabilities

In this exercise, use a normal distribution with $\mu = 69.6$ and $\sigma = 3.0$.

Method 1: To find P(X < 66), press **2ⁿᵈ** **[DISTR]** , select **2:normalcdf(** and type in **-1E99** ⏐, **66** ⏐, **69.6** ⏐, **3** ⏐) and press **ENTER**. (Note: Since no male will have a height less than 0 inches, you could use "0" in place of –1 EE 99 as your lowerbound.

```
normalcdf(-1E99,
66,69.6,3)
      .1150697316
```

Method 2: To find P(X < 66) and include a graph, you must first clear the Y-registers and turn OFF all STATPLOTS. Next, set up the Graph Window. Press **WINDOW** and set **Xmin = 69.6 - 3 * 3.0 and Xmax = 69.6 + 3 * 3.0.** Set **Xscl** = 3.0. Set **Ymax = .5 / 3.0 and Ymin = -.167/4.**

Press **2ⁿᵈ** **[DRAW]** and select **1:ClrDraw** and press **ENTER** **ENTER**. Press **2ⁿᵈ** **[DISTR]**, highlight **DRAW** and select **1:ShadeNorm(** and type in **-1E99** ⏐, **66** ⏐, **69.6** ⏐, **3**) and press **ENTER**.

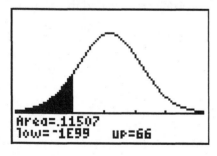

To find p(66< **X** < 72), press **2**^{**nd**} **[DISTR]** , select **2:normalcdf(** and type in **66**
, **72** **,** **69.6 ,** **3)** and press ENTER .

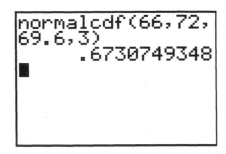

or press **2**^{**nd**} **[DRAW]** and select **1:ClrDraw** and press ENTER ENTER. Press
2^{**nd**} **[DISTR]** , highlight **DRAW** and select **1:ShadeNorm(** and type in **66** ,
72 , **69.6** , **3**) and press ENTER.

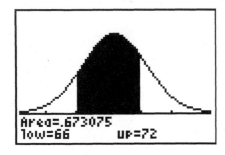

To find P(X > 72), press **2**^{**nd**} **[DISTR]** , select **2:normalcdf(** and type in **72**
, **1E99 ,** **69.6 ,** **3)** and press ENTER. (Note: In this example, the
lowerbound is 72 and the upperbound is positive infinity).

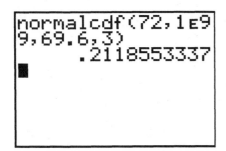

or press **2**^{**nd**} **[DRAW]** and select **1:ClrDraw** and press ENTER ENTER. Press
2^{**nd**} **[DISTR]** , highlight **DRAW** and select **1:ShadeNorm(**and type in **72** ,
1E99 , **69.6 ,** **3**) and press ENTER.

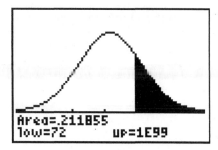

```
Area=.211855
low=72        up=1E99
```

Note: When using the TI-83/84 (or any other technology tool), the answers you obtain may vary slightly from the answers that you would obtain using the standard normal table. Consequently, your answers may not be exactly the same as the answers found in the answer key in your textbook. The differences are simply due to rounding.

Section 5.3

▶ Example 4 (pg. 264) Finding a Specific Data Value

This is called an inverse normal problem and the command **invNorm(area,** μ , σ **)** is used. In this type of problem, a percentage of the area under the normal curve is given and you are asked to find the corresponding X-value. In this example, the percentage given is the top 5 %. The TI-83/84 always calculates probability from negative infinity up to the specified X-value. To find the X-value corresponding to the top 5 %, you must accumulate the bottom 95 % of the area. Press **2ⁿᵈ [DISTR]** and select **3:invNorm(** and type in **.95** ⎵ **75** ⎵ **6.5**) and press **ENTER**.

```
invNorm(.95,75,6
.5)
        85.69154857
```

In order to score in the top 5 %, you must earn a score of at least 85.69. Assuming that scores are given as whole numbers, your score must be at least 86.

◀

▶ Exercise 40 (pg. 267) Heights of Males

This is a normal distribution with $\mu = 69.6$ and $\sigma = 3$.

a. To find the 90th percentile, press 2nd [DISTR] and select 3:invNorm(and type in .90 , 69.6 , 3) and press ENTER.

```
invNorm(.90,69.6
,3)
          73.4446547
```

b. To find the first quartile, press 2nd [DISTR] and select 3:invNorm(and type in .25 , 69.6 , 3) and press ENTER.

```
invNorm(.25,69.6
,3)
          67.57653075
```

Section 5.4

▶ Example 4 (pg. 275) Probabilities for Sampling Distributions

In this example, data has been collected on the average daily driving time for different age groups. From the graph on pg. 275, you will find that the mean driving time for adults in the 15 to 19 age group is: $\mu = 25$ minutes. The problem states that the assumed standard deviation is $\sigma = 1.5$ minutes.

You randomly sample 50 drivers in the 15 – 19 age group. Since 'n' (the sample size) is greater than 30, you can conclude that the sampling distribution of the sample mean is approximately normal with $u_{\bar{x}} = 25$ and $\sigma_{\bar{x}} = 1.5/\sqrt{50}$. To calculate P(24.7 < $\bar{x}$ < 25.5), press **2ⁿᵈ** **[DISTR]**, select **2:normalcdf(** and type in **24.7** $\boxed{,}$ **25.5** $\boxed{,}$ **25** $\boxed{,}$ **1.5/$\sqrt{50}$** $\boxed{)}$ and press **ENTER**.

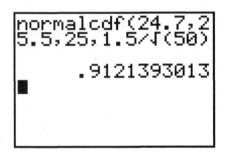

```
normalcdf(24.7,2
5.5,25,1.5/√(50)
        .9121393013
```

Note: The answer in your textbook is 0.9116. This answer was calculated using the z-table. Since z-values in the table are rounded to hundredths, the answers will vary slightly from those obtained using the TI-83/84.

◀

▶ Example 6 (pg. 277) Finding Probabilities for x and $\bar{x}$

The population is normally distributed with $\mu = 2870$ and $\sigma = 900$.

1. To calculate P(X < 2500), press 2nd [DISTR] , select 2:normalcdf(and type in **-1E99** ⎡,⎤ **2500** ⎡,⎤ **2870** ⎡,⎤ **900** ⎡)⎤ and press ENTER. (Note: Since the minimum credit card balance is 0, the lowerbound could be set at 0, rather than negative infinity.)

```
normalcdf(-1E99,
2500,2870,900)
        .3404955712
■
```

2. To calculate P($\bar{x}$ < 2500), press 2nd [DISTR] , select 2:normalcdf(and type in **-1E99** ⎡,⎤ **2500** ⎡,⎤ **2870** ⎡,⎤ **900/$\sqrt{25}$** ⎡)⎤ and press ENTER.

```
normalcdf(-1E99,
2500,2870,900/√(
25))
      .0199126205
■
```

◀

> ▸ Exercise 33 (pg. 281) Make a Decision

To decide whether the machine needs to be reset, you must decide how unlikely it would be to find a mean of 127.9 from a sample of 40 cans if, in fact, the machine is actually operating correctly at $\mu = 128$. One method of determining the likelihood of $\bar{x} = 127.9$ is to calculate how far 127.9 is from the mean of 128. You can do this by calculating how much area there is under the normal curve to the left of 127.9. The smaller that area is, the farther 127.9 is from the mean and the more unlikely 127.9 is.

To calculate P($\bar{x} \leq 127.9$), press **2ⁿᵈ** **[DISTR]** , select **2:normalcdf(** and type in   **-1E99** $\boxed{,}$ **127.9** $\boxed{,}$ **128** $\boxed{,}$ **0.20/**$\sqrt{40}$ $\boxed{)}$ and press **ENTER**.

```
normalcdf( -1E99,
127.9,128,.20/√(
40))
    7.827669991 E -4
```

Notice that the answer is displayed in scientific notation: 7.827E-4. Convert this to standard notation, .0007827, by moving the decimal point 4 places to the left. This probability is extremely small; therefore, the event ($\bar{x} \leq 127.9$) is highly unlikely if the mean is actually 128. So, something has gone wrong with the machine and the actual mean must have shifted to a value less than 128.

◀

> ▶ Technology (pg. 303) Age Distribution in the U. S.

Exercise 1: Press **STAT** and select **1:EDIT**. Clear **L1, L2** and **L3**.
Enter the age distribution into **L1** and **L2** by putting class midpoints in **L1** and
relative frequencies (converted to decimals) into **L2**. (The first entry is **2** in **L1**
and **.068** in **L2**.) To find the population mean, μ, and the population standard
deviation, σ, press **STAT**, highlight **CALC**, select **1:1-Var Stats**, press
ENTER and press **2nd** **[L1]** **,** **2nd** **[L2]** **ENTER**. The mean and the standard
deviation will be displayed. (Note: Use "σx" because the age data represents
the entire population distribution of ages, not a sample.)

Exercise 2: Enter the thirty-six sample means into **L3**. To find the mean
and standard deviation of these sample means, press **STAT**, highlight
CALC, select **1:1-Var Stats**. Press **ENTER** and press **2nd** **[L3]** **ENTER**.
The mean and the standard deviation will be displayed. (Note: use "sx"
because the 36 sample means are a sample of 36 means, not the entire
population of all possible means of size n = 40).

Exercise 3: Use the histogram on pg. 303 to answer this question.

Exercise 4: The TI-83/84 will draw a frequency histogram for a set of
data, not a *relative* frequency histogram. (The shape of the data can be
determined from either type of histogram). Press **2nd** **[STAT PLOT]** and
select **1: Plot 1.** Turn **ON** Plot 1, select **Histogram** for **Type,** set **Xlist** to
L3 and set **Freq** to **1.** Press **ZOOM** and **9** for **ZoomStat.** To adjust the
histogram so that it has nine classes, press **WINDOW**. For Xmin, use a
value slightly smaller than the minimum value of the sample means; for
Xmax, use a value slightly larger than the maximum value of the sample
means; approximate the class width by finding the range of values (max –
min) of the sample means and dividing this range by 9. This value is
approximately 2, so set **Xscl = 2** and press **GRAPH**.

Exercise 5: See the output from Exercise 1 for the population standard
deviation, σ.

Exercise 6: See the output from Exercise 2 for the standard deviation of
the 40 sample means. This standard deviation, $s_{\bar{x}}$, is an approximation
of $\sigma_{\bar{x}}$. The Central Limit Theorem states that $\sigma_{\bar{x}} \approx \dfrac{\sigma}{\sqrt{n}}$. Use your results
to confirm this fact.

Confidence Intervals

CHAPTER

6

Section 6.1

▶ Example 4 (pg. 314) Constructing a Confidence Interval

Enter the data from Example 1 on pg. 310 into **L1**. The sample standard deviation, s, for this data set is approximately 5.0. In this example, n > 30, so the sample standard deviation is a good approximation to σ, the population standard deviation. Using this sample standard deviation as an estimate of σ, you can construct a Z-Interval, a confidence interval for μ, the population mean. Press **STAT**, highlight **TESTS** and select **7:Zinterval.**

```
EDIT CALC TESTS
1:Z-Test…
2:T-Test…
3:2-SampZTest…
4:2-SampTTest…
5:1-PropZTest…
6:2-PropZTest…
7:ZInterval…
```

On the first line of the display, you have the option to select **Data** or **Stats.** For this example, select **Data** because you are using the actual data which is in **L1**. Press **ENTER**. Move to the next line and enter 5.0, the estimate of σ. On the next line, enter **L1** for **LIST**. For **Freq**, enter **1**. For **C-Level**, enter **.99** for a 99% confidence interval. Move the cursor to **Calculate.**

```
ZInterval
 Inpt:Data Stats
 σ:5
 List:L1
 Freq:1
 C-Level:99
 Calculate
```

Press **ENTER**.

```
ZInterval
 (10.579,14.221)
 x̄=12.4
 Sx=5.010193691
 n=50
```

A 99% confidence interval estimate of μ, the population mean is (10.579, 14.221). The output display includes the sample mean (12.4), the sample standard deviation (5.010), and the sample size (50).

> ▸ Example 5 (pg. 315) Confidence Interval for μ (σ known)

In this example, $\bar{x} = 22.9$ years, n = 20, a value for σ from previous studies is given as $\sigma = 1.5$ years, and the population is normally distributed. Construct a 90% confidence interval for μ, the mean age of all students currently enrolled at the college.

Press **STAT**, highlight **TESTS** and select **7:ZInterval**. In this example, you do not have the actual data. Instead you have the summary statistics of the data, so select **Stats** and press **ENTER**. Enter the value for σ : **1.5**; enter the value for $\bar{x}$: **22.9**; enter the value for **n: 20**; and enter **.90** for **C-level**. Highlight **Calculate**.

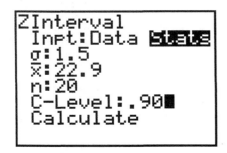

```
ZInterval
 Inpt:Data Stats
 σ:1.5
 x̄:22.9
 n:20
 C-Level:.90■
 Calculate
```

Press **ENTER**.

```
ZInterval
 (22.348,23.452)
 x̄=22.9
 n=20


■
```

Using a 90% confidence interval, you estimate that the average age of all students is between 22.348 and 23.452.

▶ Exercise 51 (pg. 319) Newspaper Reading Times

Enter the data into **L1**. Press **STAT**, highlight **TESTS** and select **7:ZInterval**.
Since you have entered the actual data points into **L1**, select **Data** for **Inpt** and
press **ENTER**. Enter the value for σ , **1.5**. Set **LIST = L1, Freq = 1** and
C-level = .90. Highlight **Calculate.**

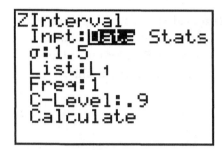

```
ZInterval
 Inpt:DATA Stats
 σ:1.5
 List:L₁
 Freq:1
 C-Level:.9
 Calculate
```

Now press **ENTER**.

```
ZInterval
 (8.4296,9.7037)
 x̄=9.066666667
 Sx=1.579632266
 n=15
■
```

Repeat the process and set **C-level = .99**.

```
ZInterval
 (8.0691,10.064)
 x̄=9.066666667
 Sx=1.579632266
 n=15
■
```

◀

▶ Exercise 65 (pg. 322) Using Technology

Enter the data into L1. In this example, no estimate of σ is given so you should use Sx, the sample standard deviation as a good approximation of σ since n > 30. Press **STAT**, highlight **CALC**, select **1:1-Var Stats** and press **L1 ENTER**. The sample standard deviation, Sx, is 13.20.

Press **STAT**, highlight **TESTS** and select **7:ZInterval**. For **Inpt**, select **Data**. For σ, enter **13.20**. Set **List** to **L1**, **Freq** to **1** and **C-level** to **.95**. Highlight **Calculate** and press **ENTER**.

```
ZInterval
 (234.13,243.42)
 x̄=238.7741935
 Sx=13.20027696
 n=31

■
```

Section 6.2

▶ Example 3 (pg.328) Constructing a Confidence Interval

In this example, n = 20, $\bar{x}$ = 6.22 and s = 0.42 and the underlying population is assumed to be normal. Find a 99% confidence interval for μ. First, notice that σ is unknown. The sample standard deviation, s, is not a good approximation of σ when n < 30. To construct the confidence interval for μ, the correct procedure under these circumstances (n < 30, σ is unknown and the population is assumed to be normally distributed) is to use a T-Interval.

Press **STAT**, highlight **TESTS**, scroll through the options and select **8:TInterval** and press **ENTER**. Select **Stats** for **Inpt** and press **ENTER**. Fill in $\bar{x}$, Sx, and n with the sample statistics. Set **C-level** to **.99**. Highlight **Calculate**.

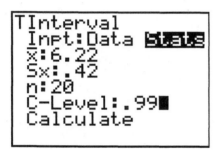

Press **ENTER**.

```
TInterval
 (5.9513,6.4887)
 x̄=6.22
 Sx=.42
 n=20
```

▶ Exercise 23 (pg. 332) Deciding on a Distribution

In this example, n = 70, $\bar{x}$ = 1.25 and s = 0.05. Notice that σ is unknown. You can use s, the sample standard deviation, as a good approximation to σ in this case because n > 30. To calculate a 95 % confidence interval for μ, press **STAT**, highlight **TESTS** and select **7:ZInterval**. Fill in the screen with the appropriate information.

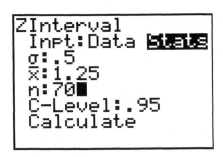

```
ZInterval
 Inpt:Data Stats
 σ:.5
 x̄:1.25
 n:70█
 C-Level:.95
 Calculate
```

Calculate the confidence interval.

```
ZInterval
 (1.1329,1.3671)
 x̄=1.25
 n=70


█
```

▸ Exercise 25 (pg. 332) Deciding on a Distribution

Enter the data into **L1**. In this case, with n = 25 and σ unknown, and assuming
that the population is normally distributed, you should use a Tinterval. Press
STAT, highlight **TESTS** and select **8:TInterval**. Fill in the appropriate
information.

```
TInterval
 Inpt:Data Stats
 List:L₁
 Freq:1
 C-Level:.95█
 Calculate
```

Calculate the confidence interval.

```
TInterval
 (20.49,23.35)
 x̄=21.92
 Sx=3.463139231
 n=25
```

Section 6.3

▶ Example 2 (pg. 336)   Constructing a Confidence Interval for p

In this example, 1219 American adults are surveyed and 354 say that their favorite sport to watch is football (see Example 1, pg. 334). Construct a 95 % confidence interval for p, the proportion of all Americans who say that their favorite sport to watch is football.

Press **STAT**, highlight **TESTS**, scroll through the options and select **A:1-PropZInt**. The value for X is the number of American adults in the group of 1219 who said that football was their favorite sport to watch, so **X = 354**. The number who were surveyed is n, so **n = 1219**. Enter **.95** for **C-level**.

```
1-PropZInt
 x:354
 n:1219█
 C-Level:.95
 Calculate
```

Highlight **Calculate** and press **ENTER**.

```
1-PropZInt
 (.26492,.31589)
 p̂=.2904019688
 n=1219


█
```

In the output display the confidence interval for p is (.26492, .31589). The sample proportion, $\hat{p}$, is .2904 and the number surveyed is 1219.

▶ **Example 3 (pg. 337)** Confidence Interval for p

From the graph, the sample proportion, $\hat{p}$, is 0.63 and n is 900. Construct a
99% confidence interval for the proportion of adults who think that teenagers are
the more dangerous drivers. In order to construct this interval using the TI-
83/84, you must have a value for X, the number of people in the study who said
that teenagers were the more dangerous drivers. Multiply 0.63 by 900 to get this
value. If this value is not a whole number, round to the nearest whole number.
For this example, X is 567.

Press **STAT**, highlight **TESTS** and select **A:1-PropZInt** by scrolling through
the options or by simply pressing **ALPHA** **A**. Enter the appropriate
information from the sample.

```
1-PropZInt
 x:567
 n:900
 C-Level:.99█
 Calculate
```

Highlight **Calculate** and press **ENTER**.

```
1-PropZInt
 (.58855,.67145)
 p̂=.63
 n=900
```

▸ Exercise 27a (pg. 341-342) Confidence Interval for p

a. To construct a 99% confidence interval for the proportion of adults from the United States who believe that human activities are contributing to global warming, begin by multiplying .65 by 2563 to obtain a value for X, the number of adults in the U.S. sample who believe that human activities are contributing to global warming. Press **STAT**, highlight **TESTS** and select **A:1-PropZInt**. Fill in the appropriate values.

```
1-PropZInt
 x:1666
 n:2563
 C-Level:.99█
 Calculate
```

Highlight **Calculate** and press **ENTER**.

```
1-PropZInt
 (.62575,.67429)
 p̂=.6500195084
 n=2563

█
```

▶ Technology (pg. 359) "Most Admired" Polls

1. Use the survey information to construct a 95% confidence interval for the proportion of people who would have chosen George W. Bush as their most admired man. Multiply .13 by 1010 to obtain a value for X, the number of people in the sample who chose George W. Bush as their most admired man.Press **STAT**, highlight **TESTS** and select **A:1-PropInt**. Enter the values for **X** and **n**. Set **C-level** to **.95**, highlight **Calculate** and press **ENTER**.

3. To construct a 95% confidence interval for the proportion of people who would have chosen Oprah Winfrey as their most admired female, you must calculate X, the number of people in the sample who chose Oprah. Multiply .09 by 1010 and round your answer to the nearest whole number. Press **STAT**, highlight **TESTS**, select **A:1-PropZInt** and fill in the appropriate information. Press **Calculate** and press **ENTER**.

4. To do one simulation, press **MATH**, highlight **PRB**, select **7:randBin(** and type in **1010** , **.12**) . The output is the number of successes in a survey of n = 1010 people. In this case, a "success" is choosing Oprah Winfrey as your most admired female. (Note: It takes approximately one minute for the TI-83/84 to do the calculation).

   To run the simulation ten times use **7:randBin(1010, .10, 10)**. (These simulations will take approximately 8 minutes). The output is a list of the number of successes in each of the 10 surveys. Calculate $\hat{p}$ for each of the surveys. $\hat{p}$ is equal to (number of successes)/1010. Use these 10 values of $\hat{p}$ to answer questions 4a and 4b.

◀

Hypothesis Testing with One Sample

CHAPTER

7

Section 7.2

▶ Example 4 (pg. 382) Hypothesis Testing Using P-values

The hypothesis test, $H_o : \mu \geq 30$ vs. $H_a : \mu < 30$, is a left-tailed test. The sample statistics are $\bar{x} = 28.5$, $s = 3.5$ and $n = 36$. The sample size is greater than 30, so the **Z-Test** is the appropriate test. To run the test, press **STAT**, highlight **TESTS** and select **1:Z-Test**. Since you are using sample statistics for the analysis, select **Stats** for **Inpt** and press **ENTER**. For μ_0 enter 30, the value for μ in the null hypothesis. For σ enter **s**, the sample standard deviation. (Note: s, the sample standard deviation, is a good approximation for σ, the population standard deviation, when n is large.) Enter **28.5** for $\bar{x}$ and **36** for **n**. On the next line, choose the appropriate alternative hypothesis and press **ENTER**. For this example, it is $< \mu_0$, a left-tailed test.

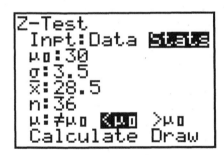

There are two choices for the output of this test. The first choice is **Calculate**. The output displays the alternative hypothesis, the calculated z-value, the P-value, $\bar{x}$ and n.

```
Z-Test
 μ<30
 z=-2.571428571
 P=.0050640239
 x̄=28.5
 n=36
```

Since p = .005, which is less than α , the correct conclusion is to **Reject** H_o .

To view the second output option, you should start by turning **OFF** any **STATPLOT** that is turned **ON**. Press **2ⁿᵈ Y=** to display the **STATPLOTS**. If a **PLOT** is **ON**, select it and move the cursor to OFF and press **ENTER**. Also, clear all Y-registers.

Press **STAT**, highlight **TESTS**, and select **1:Z-Test**. All the necessary information for this example is still stored in the calculator. Scroll down to the bottom line and select **DRAW**. A normal curve is displayed with the left-tail area of .0051 shaded. This shaded area is the area to the left of the calculated Z-value. (Because the area is so small in this example, it is not actually visible in the diagram.) The Z-value and the P-value are also displayed.

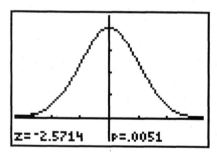

▶ Example 5 (pg. 383)   Hypothesis Testing Using P-values

This test is a two-tailed test for $H_o : \mu = 143260$ vs. $H_a : \mu \neq 143260$. The sample statistics are $\bar{x} = 135000$, s = 30000 and n = 30. Press **STAT**, highlight **TESTS** and select **1:Z-Test**. Choose **Stats** for **Inpt** and press **ENTER**. For μ_0 enter 143260, the value for μ in the null hypothesis. For σ, enter **s**, the sample standard deviation. Enter **135000** for $\bar{x}$ and **30** for **n**. On the next line, choose the appropriate alternative hypothesis and press **ENTER**. For this example, it is $\neq \mu_0$, a two-tailed test.

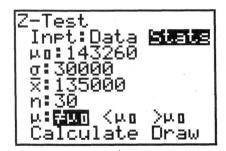

Highlight **Calculate** and press **ENTER**.

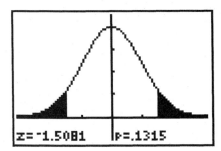

Or, highlight **Draw** and press **ENTER**.

Notice the P-value is equal to .1315. In this example, α is .05. Since the P-value is greater than α, the correct conclusion is to **Fail to Reject** H_o.

▶ Exercise 33 (pg. 391) Testing Claims Using P-values

Test the hypotheses: $H_o : \mu \le 275$ vs. $H_a : \mu > 275$. The sample statistics are $\bar{x} = 282$, s = 35 and n = 85. Press **STAT**, highlight **TESTS** and select **1:Z-Test**. For **Inpt**, choose **Stats** and press **ENTER**. Fill the input screen with the appropriate information. Choose > μ_0 for the alternative hypothesis and press **ENTER**. Highlight **Calculate** and press **ENTER**.

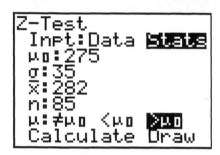

Or, highlight **Draw** and press **ENTER**.

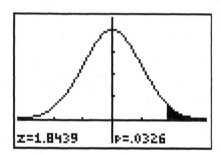

The test statistic is Z = 1.8439 and the P-value is .0326. Since the P-value is less than α, the correct conclusion is to **Reject** H_o.

▶ Exercise 37(pg. 392) Testing Claims Using P-values

This is a hypothesis test for $H_o : \mu = 15$ vs. $H_a : \mu \neq 15$. Since n > 30, the appropriate test is the Z-Test. This procedure requires a value for σ. In cases with n > 30, the sample standard deviation, s, can be used to approximate σ. Begin the analysis by entering the 32 data points into **L1**. Press **STAT**, highlight **CALC** and choose **1:1-Var Stats** and press 2nd **L1 ENTER**. The sample statistics will be displayed on the screen. The value you need for the hypothesis test is Sx, the sample standard deviation, which is 4.29.

To perform the test, press **STAT**, highlight **TESTS** and select **1:Z-Test**. Since you have the actual data points for the analysis, select **Data** for **Inpt** and press **ENTER**. Fill in the input screen with the appropriate information and select $\neq \mu_0$, as the alternative hypothesis and press **ENTER**.

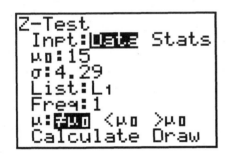

Highlight **Calculate** and press **ENTER**

```
Z-Test
 μ≠15
 z=-.2183954511
 P=.8271210783
 x̄=14.834375
 Sx=4.287612267
 n=32
■
```

Or, highlight **Draw** and press **ENTER**.

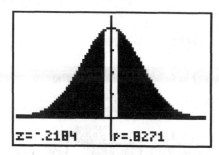

Since the P-value is greater than α, the correct conclusion is to **Fail to Reject** H_o.

▶ Exercise 43 (pg. 393) Testing Claims

This is a hypothesis test for $H_o : \mu \le 32$ vs. $H_a : \mu > 32$. Since n > 30, the appropriate test is the Z-Test. This procedure requires a value for σ. In cases with n > 30, the sample standard deviation, s, can be used to approximate σ. Begin the analysis by entering the 34 data points into **L1**. Press **STAT**, highlight **CALC** and choose **1:1-Var Stats** and press 2nd **L1** **ENTER**. The sample statistics will be displayed on the screen. The value you need for the hypothesis test is Sx, the sample standard deviation, which is 9.16.

To run the hypothesis test, press **STAT**, highlight **TESTS** and select **1:Z-Test**. Since you have the actual data points for the analysis, select **Data** for **Inpt** and press **ENTER**. Enter **32** for μ_0 and enter **9.16** for σ. The data is stored in **L1** and **Freq** is **1**. The alternate hypothesis is a right-tailed test so select $> \mu_0$ and press **ENTER**.

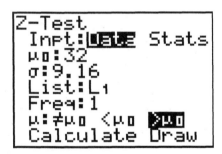

To run the test, select **Calculate** and press **ENTER**.

```
Z-Test
 μ>32
 z=-1.479081684
 p=.9304407244
 x̄=29.67647059
 Sx=9.164227515
 n=34
```

Or, select **Draw** and press **ENTER**.

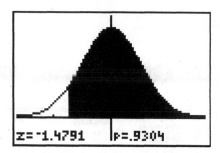

Since the P-value of .9304 is greater than α, the correct conclusion is to **Fail to Reject** H_o.

Section 7.3

▶ Example 6 (pg. 402)   Using P-values with a T-Test

This test is a two-tailed test of $H_o: \mu = 118$ vs. $H_a: \mu \neq 118$. The sample statistics are $\bar{x} = 128$, s = 20 and n = 11. Since n < 30, the test is a T-Test if you assume that the underlying population is approximately normally distributed. Press **STAT**, highlight **TESTS** and select **2:T-Test**. Choose **Stats** for **Inpt** and press **ENTER**. Fill in the following information: $\mu_0 = \textbf{118}$, $\bar{x} = \textbf{128}$, **Sx = 20** and **n = 11**. Choose the two-tailed alternative hypothesis, $\neq \mu_0$, and press **ENTER**. Highlight **Calculate** and press **ENTER**

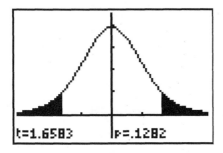

Or, highlight **Draw** and press **ENTER**.

Since the P-value is greater than α, the correct conclusion is to **Fail to Reject** H_o.

▶ Exercise 29 (pg. 405) Testing Claims Using P-values

The correct hypothesis test is $H_o : \mu = 3$ vs. $H_a : \mu < 3$. Enter the data into L1. Since n < 30, the appropriate test is the T-Test, if you assume that the underlying population is approximately normal. Press **STAT**, highlight **TESTS** and select **2:T-Test**. Select **Data** for **Inpt** and press **ENTER**. Fill in the screen with the necessary information, choose **Calculate** and press **ENTER**.

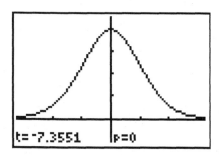

```
T-Test
 μ<3
 t=-7.355109012
 P=2.8526838E-7
 x̄=1.925
 Sx=.6536335687
 n=20
■
```

The p-value is so small that it is written in scientific notation: 2.8526838 E -7, which is equal to 0.00000028526838, or approximately 0.

Or, choose **Draw** and press **ENTER**.

```
t=-7.3551    │P=0
```

Since the P-value is less than α , the correct conclusion is to **Reject** H_o .

◀

▸ Exercise 37(pg. 406) Deciding on a Distribution

This test is a left-tailed test of $H_o : \mu \geq 23$ vs. $H_a : \mu < 23$. The sample statistics are $\bar{x} = 22$, s = 4 and n = 5. Since n < 30 and gas mileage is normally distributed, the appropriate test is the **T-Test**. Press **STAT**, highlight **TESTS** and select **2:T-Test**. Fill in the screen with the necessary information and choose **Calculate** and press **ENTER**.

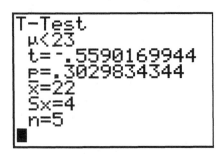

Or, choose **Draw** and press **ENTER**.

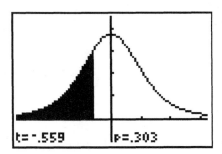

Since the P-value is greater than α, the correct conclusion is to **Fail to Reject** H_o.

▶ Exercise 38 (pg. 406) Deciding on a Distribution

This test is a two-tailed test of $H_o : \mu = \$23,000$ vs. $H_a : \mu < \$23,000$. The sample statistics are $\bar{x} = \$21,856$, s = \$3163$ and n = 50. Since n $\geq$30, s is a good approximation to σ, so the appropriate test is the **Z-Test**. Press **STAT**, highlight **TESTS** and select **1:Z-Test**. Fill in the screen with the necessary information and choose **Calculate** and press **ENTER**.

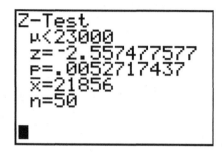

Or, choose **Draw** and press **ENTER**.

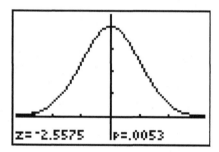

Since the P-value is smaller than α, the correct conclusion is to **Reject** H_o.

Section 7.4

> **Example 1 (pg. 408)**   Hypothesis Test for a Proportion

This hypothesis test is a left-tailed test of: $H_o : p \geq .20$ vs. $H_a : p < .20$. The first step is to verify that the products 'np' and 'nq' are both great than 5. Since both products are greater than 5, the Z-test is the appropriate test. The sample statistics are $\hat{p} = .15$ and n = 100. Press **STAT**, highlight **TESTS** and select **5:1-PropZTest**. This test requires a value for p_0, which is the value for p in the null hypothesis. Enter **.20** for p_0. Next, a value for X is required. X is the number of "successes" in the sample. In this example, a success is " having a wireless network at home." Since 15% of the individuals in the sample say that they have a wireless network at home, **X** is equal to .15 times 100 or **15**. Next, enter the value for n. Select $< p_0$ for the alternative hypothesis and press **ENTER**.

```
1-PropZTest
 p0:.2
 x:15
 n:100
 prop≠p0  <p0  >p0
 Calculate Draw
```

Highlight **Calculate** and press **ENTER**.

```
1-PropZTest
 prop<.2
 z=-1.25
 p=.105649839
 p̂=.15
 n=100
```

Or, highlight **Draw** and press **ENTER**.

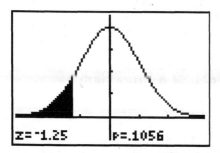

Since the P-value is greater than α, the correct conclusion is to **Fail to Reject** H_o.

▶ Example 3 (pg. 410)   Hypothesis Test for a Proportion

First, verify that the products 'np' and 'nq' are both greater than 5. The hypothesis test is: $H_o : p \leq .55$ vs. $H_a : p > .55$. The sample statistics are n = 425 and X = 255. Press **STAT**, highlight **TESTS** and select **5:1-PropZTest**. Enter the necessary information. Highlight **Calculate** and press **ENTER**.

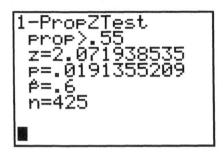

Or, highlight **Draw** and press **ENTER**.

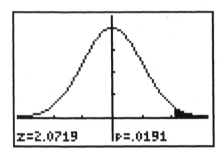

Since the P-value is less than α, the correct conclusion is to **Reject** H_o.

▸ Exercise 11 (pg. 411) Testing Claims

Use a right-tailed hypothesis test to test the hypotheses: $H_o : p \le .30$ vs. $H_a : p > .30$. The sample statistics are n = 1050 and $\dot{p}$ = .32. Multiply n times $\hat{p}$ to find X, the number of consumers in the sample who have stopped purchasing a product because it pollutes the environment. Press **STAT**, highlight **TESTS** and select **5:1-PropZTest**. Enter the necessary information.

Highlight **Calculate** and press **ENTER**.

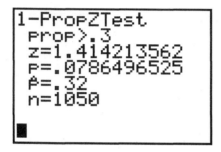

Or, highlight **Draw** and press **ENTER**.

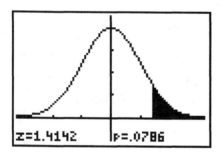

Since the P-value is .greater than α, the correct conclusion is to **Fail to Reject** H_o.

Section 7.5

▸ Example 4 (pg. 417) Hypothesis Test for a Variance

This hypothesis test is a right-tailed test of: $H_o : \sigma^2 \leq 0.25$ vs. $H_a : \sigma^2 > 0.25$. The sample statistics are $s^2 = 0.27$ and $n = 41$. First you must calculate the χ^2 value: $((41 - 1) *0.27)/0.25 = 43.2$. Next, find the p-value associated with this χ^2 value. Press 2nd **DISTR**, and scroll down to χ^2 cdf(. This function requires a *lowerbound*, an *upperbound* and the *degrees of freedom*. Since this is a right-tailed test, 43.2 is the *lowerbound* and positive infinity (1E99) is the *upperbound*. The degrees of freedom (n-1) is equal to 40.

```
X²cdf(43.2,1E99,
40)
        .3362231333
■
```

The P-value is 0.336. Since this is greater than α, the correct conclusion is to **Fail to Reject** H_o.

◀

▶ Technology (pg. 433) The Case of the Vanishing Women

Exercise 1: Use the TI-83/84 to run the hypothesis test and compare your results to the MINITAB results shown in the display at the bottom of pg. 433. Use a two-tailed test to test the hypotheses: $H_o : p = 0.53$ vs. $H_a : p \neq 0.53$. The sample statistics are X = 102 women and n = 350 people selected from the Boston City Directory. Press **STAT**, highlight **TESTS** and select **5:1-PropZTest**. Fill in the appropriate information highlight **Calculate** and press **ENTER** or, highlight **Draw** and press **ENTER**.

Exercise 4: In the first stage of the jury selection process, 350 people are selected and 102 of them are women. So, at this stage, the proportion of women is 102 out of 350, or 0.2914. From this population of 350 people, a sample of 100 people is selected at random and only nine are women. Test the claim that the proportion of women in the population is 0.2914. Use a two-tailed test to test the hypotheses:

$H_o : p = 0.2914$ vs. $H_a : p \neq 0.2914$. The sample statistics are X = 9 women and n = 100 people. Press **STAT**, highlight **TESTS** and select **5:1-PropZTest**. Fill in the appropriate information, Highlight **Calculate** and press **ENTER** or highlight **Draw** and press **ENTER**.

◀

Hypothesis Testing with Two Samples

CHAPTER

8

Section 8.1

▶ Example 3 (pg. 443) Two-Sample Z-Test

Test the claim that the average daily cost for meals and lodging when vacationing in Texas is less than the average costs when vacationing in Virginia. Designate Texas as population 1 and Virginia as population 2. The appropriate hypothesis test is a left-tailed test of H_o: $\mu_1 \geq \mu_2$ vs H_a: $\mu_1 < \mu_2$. The sample statistics are displayed in the table at the top of pg. 443. Each sample size is greater than 30, so the correct test is a Two-sample Z-Test.

Press **STAT**, highlight **TESTS** and select **3:2-SampZTest**. Since you are using the sample statistics for your analysis, select **Stats** for **Inpt** and press **ENTER**. In order to use this test, values for σ_1 and σ_2 are required. The sample standard deviations, s_1 and s_2 can be used as approximations for σ_1 and σ_2 when both n_1 and n_2 are greater than or equal to 30. Enter 15 for σ_1. Enter 22 for σ_2. Next, enter the mean and sample size for group 1: $\bar{x}_1 = 248$ and $n_1 = 50$. Continue by entering the mean and sample size for group 2: $\bar{x}_2 = 252$ and $n_2 = 35$.

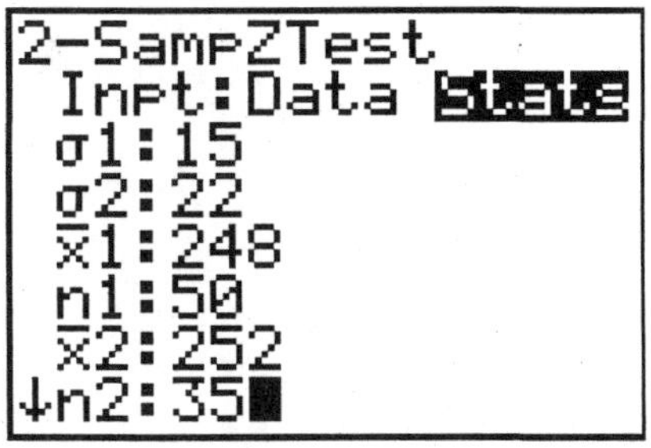

```
2-SampZTest
 Inpt:Data Stats
 σ1:15
 σ2:22
 x̄1:248
 n1:50
 x̄2:252
↓n2:35■
```

Use the down arrow to display the next line. This line displays the three possible alternative hypotheses for testing: $\mu_1 \neq \mu_2$, $\mu_1 > \mu_2$, or $\mu_1 < \mu_2$. For this example, select $< \mu_2$ and press **ENTER**. Scroll down to the next line and select **Calculate** or **Draw**.

The output for **Calculate** is displayed on two pages. (Notice that the only piece of information on the second page is n_2, so that page is not displayed here.)

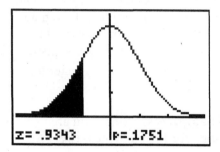

```
2-SampZTest
 μ₁<μ₂
 z=-.9343200811
 P=.1750693839
 x̄₁=248
 x̄₂=252
↓n₁=50
```

The output for **Draw** contains a graph of the normal curve with the area associated with the test statistic shaded.

```
z=-.9343        P=.1751
```

Both outputs display the P-value, which is .1751. Since the P-value is greater than α, the correct decision is to **Fail to Reject** H_o. At the 1% level of significance, there is not enough evidence to support the claim that the average cost for meals and lodging when vacationing in Texas is less than the average cost when vacationing in Virginia.

◄

▶ Exercise 19(pg. 446) Testing the Difference between Two Means

The appropriate hypothesis test is a two-tailed test of $H_o: \mu_1 = \mu_2$ vs $H_a: \mu_1 \neq \mu_2$. Designate Tire Type A as population 1 and Tire Type B as population 2. Press **STAT**, highlight **TESTS** and select **3:2-SampZTest**. For **Inpt,** select **Stats** and press **ENTER**. Enter the sample standard deviations as approximations of σ_1 and σ_2. Next, enter the sample statistics for each group.

```
2-SampZTest
 Inpt:Data ▮▮▮▮
 σ1:4.7
 σ2:4.3
 x̄1:42
 n1:35
 x̄2:45
↓n2:35
```

Select $\neq \mu_2$ as the alternative hypothesis and press **ENTER**. Scroll down to the next line and select **Calculate** or **Draw** and press **ENTER**.

The output for **Calculate** is:

```
2-SampZTest
 μ1≠μ2
 z=-2.786116393
 P=.0053344895
 x̄1=42
 x̄2=45
↓n1=35
```

The output for **Draw** is:

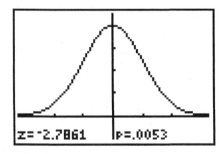

```
z=-2.7861    P=.0053
```

The test statistic, z, is -2.786 and the p-value is .0053. Since the P-value is less than α, the correct decision is to **Reject** H_o. ◀

> ▶ Exercise 29 (pg. 447) Testing the Difference between 2 Means

Enter the data from 1981 into **L1** and enter the data from the more recent study into **L2**. The hypothesis test is: $H_o: \mu_1 \le \mu_2$ vs $H_a: \mu_1 > \mu_2$. Since both n_1 and n_2 equal 30, the Two-Sample Z-Test can be used. In order to use this test, values for σ_1 and σ_2 are required. The sample standard deviations, s_1 and s_2 can be used as approximations for σ_1 and σ_2 when both n_1 and n_2 are greater than or equal to 30. To find the standard deviations, press **STAT**, highlight **CALC** and select **1:1-Var Stats** and press **2nd L1 ENTER**. The value for Sx is the sample standard deviation for the 1981 data. Next, press **STAT**, highlight **CALC** and select **1:1-Var Stats**, press **2nd L2 ENTER**. The value for Sx is the sample standard deviation for the more recent data.

To run the hypothesis, press **STAT**, highlight **TESTS** and select **3:2-SampZTest**. For **Inpt** select **Data** and press **ENTER**. Enter the sample standard deviations as approximations of σ_1 and σ_2. For **List1**, press **2nd [L1]** and for **List2** press **2nd [L2]** .

```
2-SampZTest
 Inpt:DATA Stats
 σ1:.49
 σ2:.47
 List1:L₁
 List2:L₂
 Freq1:1
↓Freq2:1
```

Select $>\mu_2$ as the alternative hypothesis and press **ENTER**. Scroll down to the next line and select **Calculate** or **Draw** and press **ENTER**.

The output for **Calculate** is:

```
2-SampZTest
 μ₁>μ₂
 z=3.011666878
 p=.001299157
 x̄1=2.13
 x̄2=1.756666667
↓Sx1=.490003519
```

The output for **Draw** is:

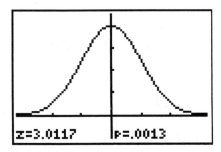

Since the P-value is smaller than α, the correct decision is to **Reject** H_o.

▶ Exercise 39 (pg. 450) Confidence Intervals for $\mu_1 - \mu_2$

Construct a confidence interval for $\mu_1 - \mu_2$, the difference between two population means. Designate the group using the DASH diet as population 1 and the group using the tradition diet as population 2. The sample statistics for group 1 are: $\bar{x}_1 = 123.1$ mm Hg, $s_1 = 9.9$ mm Hg and $n_1 = 269$. The sample statistics for group 2 are: $\bar{x}_2 = 125$ mm Hg, $s_2 = 10.1$ mm Hg and $n_2 = 268$.

Press **STAT**, highlight **TESTS** and select **9:2-SampZInt**. Since you are using the sample statistics for your analysis, select **Stats** for **Inpt** and press **ENTER**. Enter the sample standard deviations as approximations for σ_1 and σ_2. Enter the sample means and sample sizes for each group.

```
2-SampZInt
 Inpt:Data Stats
 σ1:9.9
 σ2:10.1
 x̄1:123.1
 n1:269
 x̄2:125
↓n2:268
```

Scroll down to the next line and type in the confidence level of .95. Scroll down to the next line and press **ENTER**.

```
2-SampZInt
 (-3.592,-.2083)
 x̄1=123.1
 x̄2=125
 n1=269
 n2=268
```

A 95 % confidence interval for the difference in the population means is (-3.592, -0.2083). The average systolic blood pressure for individuals on the DASH diet is between 0.2083 and 3.592 mm Hg GT lower than the average systolic blood pressure for individuals on the traditional diet and exercise plan.

◀

Section 8.2

> **Example 1 (pg. 454)** Two Sample t-Test

To test whether the mean braking distances are different, use a two-tailed test: $H_o: \mu_1 = \mu_2$ vs $H_a: \mu_1 \neq \mu_2$. The sample statistics are found in the table at the top of pg. 454 in your textbook.

Press **STAT**, highlight **TESTS**, and select **4:2-SampTTest**. Since you are inputting the sample statistics, select **Stats** and press **ENTER**. Enter the sample information from the two samples. Select $\neq \mu_2$ as the alternative hypothesis and press **ENTER**. Scroll down to the next line. On this line, select **NO** because the variances are NOT assumed to be equal and therefore, you do not want a pooled estimate of the standard deviation. Press **ENTER**. Scroll down to the next line, highlight **Calculate** and press **ENTER**.

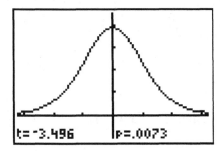

```
2-SampTTest
 µ1≠µ2
 t=-3.496035488
 p=.0072653117
 df=8.594325237
 x̄1=134
↓x̄2=143
```

The output (shown above) for **Calculate** displays the alternative hypothesis, the test statistic, the P-value, the degrees of freedom and the sample statistics. Notice the degrees of freedom = 8.594. In cases, such as this one, in which the population variances are not assumed to be equal, the calculator calculates an adjusted degrees of freedom, rather than using the smaller of (n_1-1) and (n_2-1).

If you choose **Draw**, the output includes a graph with the area associated with the P-value shaded.

▶ Example 2 (pg. 455) Two Sample t-Test

To test the claim that the mean calling range of the manufacturer's cordless phone is greater than the mean range of the competitor's phone, use a right-tailed test with $H_o: \mu_1 \leq \mu_2$ vs $H_a: \mu_1 > \mu_2$. Designate the manufacturer's data as population 1 and the competitor's data as population 2. Use the Two-Sample T-Test for the analysis, and use the pooled variance option because the population variances are assumed to be equal in this example. The sample statistics are found in the table at the top of pg. 455 in your textbook.

Press **STAT**, highlight **TESTS**, and select **4:2-SampTTest**. Fill in the input screen with the sample statistics. Select $> \mu_2$ for the alternative hypothesis and press **ENTER**. Select **YES** for **Pooled** and press **ENTER**. To do the analysis, select **Calculate** and press **ENTER**.

```
2-SampTTest
 µ1 >µ2
 t=1.811358919
 P=.0404131295
 df=28
 x̄1=1275
↓x̄2=1250
■
```

```
2-SampTTest
 µ1 >µ2
↑Sx1=45
 Sx2=30
 Sxp=37.7136769
 n1=14
 n2=16
■
```

The two-page output display (shown above) includes the alternative hypothesis, the test statistic, the P-value, degrees of freedom, the sample statistics and the pooled standard deviation (Sxp = 37.7)

If you select **Draw**, the display includes a graph with the shaded area associated with the test statistic, the P-value and the test statistic.

Since the P-value is less than α, the correct decision is to **Reject** H_o.

▸ Exercise 15 (pg. 457) Testing the Difference between 2 Means

Designate the small cars as population 1 and the midsize cars as population 2. Test the hypothesis: $H_o: \mu_1 = \mu_2$ vs. $H_a: \mu_1 \neq \mu_2$. The sample statistics are: $\bar{x}_1 = 10.1$, $s_1 = 4.11$, $n_1 = 12$, $\bar{x}_2 = 8.3$, $s_2 = 4.02$ and $n_2 = 17$. Use the Two-Sample T-test with the pooled variance option. Press **STAT**, highlight **TESTS** and select **4:2-SampTTest**. Choose **Stats** for **Inpt** and press **ENTER**. Enter the sample statistics. Select $\neq \mu_2$ as the alternative hypothesis and press **ENTER**. Select **YES** for **Pooled** and press **ENTER**. Highlight **Calculate** and press **ENTER**.

```
2-SampTTest
 µ1≠µ2
 t=1.176775083
 p=.2495559218
 df=27
 x̄1=10.1
↓x̄2=8.3
■
```

Notice that the test statistic = 1.177, and the P-value is .2496. Since the P-value is greater than α, the correct decision is to **Fail to Reject** H_o.

◀

▶ Exercise 23(pg. 459) Testing the difference between 2 Means

Designate the "Old curriculum" as population 1 and the "New curriculum" as population 2 and test the hypothesis: $H_o : \mu_1 \geq \mu_2$ vs. $H_a : \mu_1 < \mu_2$. Press **STAT**, select **1:Edit** and enter the data sets into **L1** (Old curriculum) and **L2** (New curriculum). Press **STAT**, highlight **TESTS** and select **4:2-SampTTest**. For this analysis, you are using the actual data so select **Data** for **Inpt** and press **ENTER**. Fill in the input screen with the appropriate information. Choose $< \mu_2$ as the alternative hypothesis and press **ENTER**. Choose **YES** for **Pooled** and press **ENTER**. Highlight **Calculate** and press **ENTER**.

```
2-SampTTest
 µ1<µ2
 t=-4.29519297
 P=5.0503069ᴇ-5
 df=42
 x̄1=56.68421053
↓x̄2=67.4
```

Since the P-value (.00005) is less than α, the correct decision is to **Reject** H_o. This means that the data supports H_a, indicating that the new method of teaching reading produces higher reading test scores than the old method. The recommendation is to change to the new method.

◀

▶ Exercise 25 (pg. 460) Confidence Intervals

Compare the mean calorie content of grilled chicken sandwiches from Burger King restaurants to similar chicken sandwiches from McDonald's using a confidence interval. For this exercise, assume that the populations are normal and the variances are equal. Also, notice that the sample sizes are both less than 30. The appropriate confidence interval technique is the Two-Sample t-interval.

Press **STAT**, highlight **TESTS** and select **0:2-SampTInt**. For **Inpt**, select **Stats** and press **ENTER**. Enter the sample statistics from the two sets of data. Designate Burger King as population 1 and McDonald's as population 2. Choose **.95** for **C-level** and choose **YES** for **Pooled** and press **ENTER**. Scroll down to the next line and press **ENTER** to **Calculate** the confidence interval.

```
2-SampTInt
 (24.337,35.663)
 df=25
 x̄₁=450
 x̄₂=420
 Sx₁=6.2
↓Sx₂=8.1
```

The output displays the 95 % confidence interval for ($\mu_1 - \mu_2$) and the sample statistics. One way to interpret the confidence interval is to state that "based on a 95% confidence interval, the difference in mean calorie content of grilled chicken sandwiches at the two restaurants is between 24.337 and 35.663 calories." This means that the mean calorie content of the Burger King sandwiches is anywhere from 24.337 to 35.663 calories more than that of McDonald's.

◀

▸ Exercise 27 (pg. 460) Confidence Intervals

Compare the mean cholesterol content of grilled sandwiches from Burger King to similar sandwiches from McDonald's using a confidence interval. For this exercise, assume the populations are normal but do NOT assume equal variances.

Press **STAT**, highlight **TESTS** and select **0:2-SampTInt**. For **Inpt**, select **Stats** and press **ENTER**. Enter the sample statistics from the table displayed in Exercise 27. Designate Burger King as population 1 and McDonald's as population 2. Choose **.90** for **C-level** and choose **NO** for **Pooled** and press **ENTER**. Scroll down to the next line and press **ENTER** to **Calculate** the confidence interval.

```
2-SampTInt
 (3.1678,6.8322)
 df=24.61579151
 x̄1=75
 x̄2=70
 Sx1=3.64
↓Sx2=2.12
■
```

The output displays the 90 % confidence interval for (μ_1 - μ_2), the adjusted degrees of freedom and the sample statistics. The confidence interval states that the mean cholesterol content for Burger King's sandwiches is anywhere from 3.17 to 6.83 mgs. more than that of McDonald's.

◀

Section 8.3

▶ Example 1 (pg. 463)   The Paired t-Test

In this example, the data is paired data, with two scores for each of the 8 golfers. Enter the scores that each golfer gave as his or her most recent score using the old design clubs into L1. Enter the scores achieved after using the newly designed clubs into L2. Next, you must create a set of differences, d = (old score) - (new score). To create this set, move the cursor to highlight the label **L3,** found at the top of the third column, and press **ENTER**. Notice that the cursor is flashing on the bottom line of the display. Press 2nd **[L1]** - 2nd **[L2]**

```
L1       L2       L3       3
 89       83      ------
 84       83
 96       92
 82       84
 74       76
 92       91
 85       80
L3 =L1−L2
```

and press **ENTER**.

```
L1       L2       L3       3
 89       83      6
 84       83      1
 96       92      4
 82       84      -2
 74       76      -2
 92       91      1
 85       80      5
L3(1)=6
```

Each value in **L3** is the difference **L1 - L2**.

To test the claim that golfers can lower their scores using the manufacturer's newly designed clubs, the hypothesis test is: $H_o: \mu_d \leq 0$ vs. $H_a: \mu_d > 0$. Press **STAT**, highlight **TESTS** and select **2:T-Test**. In this example, you are using the actual data to do the analysis, so select **Data** for **Inpt** and press **ENTER**. The value for μ_o is **0**, the value in the null hypothesis. The set of differences is found in **L3**, so set **List** to L3. Set **Freq** equal to **1**. Choose $> \mu_o$ as the alternative hypothesis and highlight **Calculate** and press **ENTER**.

```
T-Test
 μ>0
 t=1.498259585
 P=.0888692418
 x̄=1.625
 Sx=3.067688753
 n=8
```

You can also highlight **DRAW** and press **ENTER**.

Since the P-value is less than α, the correct decision is to **Reject** H_o.

▶ Exercise 9 (pg. 466) Paired Difference Test

Press **STAT** and select **1:Edit.** Enter the students' scores on the first SAT into **L1** and their scores on the second Sat into **L2**. Since this is paired data, create a column of differences, d = (first SAT) - (second SAT). To create the differences, move the cursor to highlight the label **L3,** found at the top of the third column, and press **ENTER**. Notice that the cursor is flashing on the bottom line of the display. Press 2nd **[L1]** - 2nd **[L2]** and press **ENTER**.

L1	L2	L3	3
445	446	-1	
510	571	-61	
429	517	-88	
452	478	-26	
629	610	19	
433	453	-20	
551	516	35	

L3(1)= -1

Each value in **L3** is the difference **L1 - L2**.

To test the claim that students' verbal SAT scores improved the second time they took the verbal SAT, the hypotheses are:: $H_o: \mu_d \geq 0$ vs. $H_a: \mu_d < 0$. Press **STAT**, highlight **TESTS** and select **2:T-Test**. In this example, you are using the actual data to do the analysis, so select **Data** for **Inpt** and press **ENTER**. The value for μ_o is **0**, the value in the null hypothesis. The set of differences is found in **L3**, so set **List** to L3. Set **Freq** equal to **1**. Choose $< \mu_o$ as the alternative hypothesis and highlight **Calculate** and press **ENTER**.

```
T-Test
 μ<0
 t=-3.001096523
 p=.0051086649
 x̄=-33.71428571
 Sx=42.03373841
 n=14
■
```

Or, highlight **DRAW** and press **ENTER**.

Since the P-value is less than α, the correct decision is to **Reject** H_o.

▶ Exercise 21 (pg. 470)   Confidence Interval for μ_d

Press **STAT** and select **1:Edit**. Enter the data for "hours of sleep without the drug" into **L1** and the data for "hours of sleep using the new drug" into **L2**. Since this is paired data, create a column of differences, **d = L1 - L2**. To create the differences, move the cursor to highlight the label **L3**, found at the top of the third column, and press **ENTER**. Notice that the cursor is flashing on the bottom line of the display. Press **2nd [L1]** - **2nd [L2]** and press **ENTER**. Each value in **L3** is the difference **L1 - L2**.

To construct a confidence interval for μ_d, press **STAT**, highlight **TESTS** and select **8:Tinterval**. For **Inpt**, select **Data**, for **List**, enter **L3**. Make sure that **Freq** is equal to **1**, and the **C-level** is **90**. Highlight **Calculate** and press **ENTER**.

```
TInterval
 (-1.763,-1.287)
 x̄=-1.525
 Sx=.5422176685
 n=16
```

The confidence interval for μ_d is -1.763 to -1.287. This means that the average difference in hours of sleep is 1.287 hours to 1.763 hours more for patients using the new drug.

◀

Section 8.4

▶ Example 1 (pg. 473) Testing the Difference Between p_1 and p_2

To test the claim that there is a difference in the proportion of female Internet users who plan to shop On-line and the proportion of male Internet users who plan to shop On-line, the correct hypothesis test is: $H_o : p_1 = p_2$ vs $H_a : p_1 \neq p_2$. Designate the females as population 1 and the males as population 2. The sample statistics are $n_1 = 200$, $\hat{p}_1 = .30$, $n_2 = 250$, and $\hat{p}_2 = .38$. To conduct this test using the TI-83/84, you need values for x_1, the number of females in the sample who plan to shop On-line, and x_2, the number of males who plan to shop On-line. To calculate x_1, multiply n_1 times $\hat{p}_1$. To calculate x_2, multiply n_2 times $\hat{p}_2$. (Note: These two values, x_1 and x_2, are given in the table on pg. 473.)

Press **STAT**, highlight **TESTS** and select **6:2-PropZTest** and fill in the appropriate information. Highlight **Calculate** and press **ENTER**.

```
2-PropZTest
 P1≠P2
 z=-1.774615984
 P=.0759612188
 P̂1=.3
 P̂2=.38
↓P̂=.344444444
```

The output displays the alternative hypothesis, the test statistic, the P-value, the sample statistics and the weighted estimate of the population proportion, $\hat{p}$. Note: At this point in the analysis, you should confirm that the following products are each equal to 5 or more: $n_1\bar{p}$, $n_1\bar{q}$, $n_2\bar{p}$ and $n_2\bar{q}$. The value for $\bar{p}$ can be found in the above output, $\bar{p} = \hat{p}$. And $\bar{q} = 1 - \bar{p}$.

Since the P-value is less than α, the correct decision is to **Reject** H_o.

> ▸ Exercise 7(pg. 475) The Difference Between Two Proportions

Designate the 1991 data as population 1 and the more recent data as population 2 and test the hypotheses: $H_o: p_1 = p_2$ vs $H_a: p_1 \neq p_2$. The sample statistics are $n_1 = 1539$, $x_1 = 520$, $n_2 = 2055$, and $x_2 = 865$.

Press **STAT**, highlight **TESTS** and select **6:2-PropZTest** and fill in the appropriate information. Highlight **Calculate** and press **ENTER**.

```
2-PropZTest
 P1≠P2
 z=-5.06166817
 P=4.1630154E-7
 p̂1=.3378817414
 p̂2=.4209245742
↓p̂=.3853644964
■
```

Note: At this point in the analysis, you should confirm that the following products are each equal to 5 or more: $n_1\bar{p}$, $n_1\bar{q}$, $n_2\bar{p}$ and $n_2\bar{q}$. The value for $\bar{p}$ can be found in the above output, $\bar{p} = \hat{p}$. And $\bar{q} = 1 - \bar{p}$.

Since the P-value (.0000004163) is less than α, the correct decision is to **Reject** H_o. This indicates that the proportion of adults using alternative medicines has changed since 1991.

◀

▶ Exercise 25 (pg. 478)   Confidence Interval for $p_1 - p_2$.

Construct a confidence interval to compare the proportion of students who had planned to study engineering several years ago to the proportion currently planning on studying engineering. Designate the earlier survey results as population 1 and the recent survey results as population 2. The sample statistics are $n_1 = 1,068,000$, $\hat{p}_1 = .088$, $n_2 = 1,476,000$, and $\hat{p}_2 = .083$. To calculate x_1 multiply n_1 times $\hat{p}_1$. To calculate x_2, multiply n_2 times $\hat{p}_2$.

Press **STAT**, highlight **TESTS** and select **B:2-PropZInt** and fill in the appropriate information. Highlight **Calculate** and press **ENTER**.

```
2-PropZInt
 (.0043,.0057)
 p̂1=.088
 p̂2=.083
 n1=1068000
 n2=1476000
```

The confidence interval (.0043, .0057) indicates that the proportion of students having chosen engineering in the past is between 0.43 % and 0.57 % higher than the proportion of students currently choosing engineering. Notice how narrow the confidence interval is. This is due to the very large sample sizes.

◀

▶ Technology (pg. 487) Tails Over Heads

Exercise 1 - 2: Test the hypotheses: H_o : P(Heads) = .5 vs. H_a : P(Heads) ≠ .5 using the one sample test of a proportion. Press **STAT**, highlight **TESTS** and select **5:1-PropZTest.** For this example, p_o = .5. Using Casey's data, X = 5772 and n = 11902. The alternative hypothesis is ≠ . Highlight **Calculate** and press **ENTER**.

```
1-PropZTest
 prop≠.5
 z=-3.281504874
 p=.0010326665
 p̂=.4849605108
 n=11902
■
```

Since the P-value is less than α , the correct decision is to **Reject** H_o .

Exercise 3: The histogram at the top of the page is a graph of 500 simulations of Casey's experiment. Each simulation represents 11902 flips of a fair coin. The bars of the histogram represent frequencies. Use the histogram to estimate how often 5772 or fewer heads occurred.

To simulate this experiment, you must use an alternative technology. The TI-83/84 does not have the memory capacity to do this experiment.

Exercise 4: To compare the mint dates of the coins, run the hypothesis test: H_o : $\mu_1 = \mu_2$ vs. H_a : $\mu_1 \neq \mu_2$. Designate the Philadelphia data as population 1 and the Denver data as population 2.

Press **STAT**, highlight **TESTS** and select **3:2-SampZTest**. Choose **Stats** for **Inpt** and press **ENTER**. Enter the sample statistics. Use the sample standard deviations as approximations to the population standard deviations. Select ≠ μ_2 as the alternative hypothesis and press **ENTER**. Highlight **Calculate** and press **ENTER**.

```
2-SampZTest
  µ₁≠µ₂
  z=8.801919011
  P=1.363493ε-18
  x̄₁=1984.8
  x̄₂=1983.4
↓n₁=7133
```

Since the P-value is extremely small, the correct decision is to **Reject** H_o.

Exercise 5: To compare the average mint value of coins found in Philadelphia to those found in Denver, run the hypothesis test: $H_o: \mu_1 = \mu_2$ vs. $H_a: \mu_1 \neq \mu_2$. Designate the Philadelphia data as population 1 and the Denver data as population 2.

Press **STAT**, highlight **TESTS** and select **3:2-SampZTest**. Choose **Stats** for **Inpt** and press **ENTER**. Enter the sample statistics. Use the sample standard deviations as approximations to the population standard deviations. Select $\neq \mu_2$ as the alternative hypothesis and press **ENTER**. Highlight **Calculate** and press **ENTER**. Since the P-value is extremely small, the correct decision is to **Reject** H_o.

Correlation and Regression

CHAPTER

9

Section 9.1

▶ Example 3 (pg. 498) Constructing a Scatter plot

Press **STAT**, highlight **1:Edit** and clear **L1** and **L2**. Enter the X-values into **L1** and the Y-values into **L2**. To prepare to construct the scatterplot, press the **Y=** key and clear all the Y-registers. Press **2ⁿᵈ** **[STAT PLOT]** , select **1:Plot1**, turn **ON** Plot 1 and press **ENTER**. For **Type** of graph, select the **scatter plot** which is the first selection. Press **ENTER**. Enter **L1** for **Xlist** and **L2** for **Ylist**. Highlight the first selection, the small square, for the type of **Mark**. Press **ENTER**. Press **ZOOM** and **9** to select **ZoomStat**.

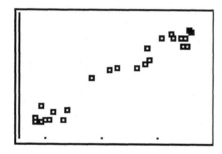

This graph shows a positive linear correlation.

▸ Example 5 (pg. 501)   Finding a Correlation Coefficient

For this example, use the data from Example 3 on pg. 498. Enter the X-values
into **L1** and the Y-values into **L2**. In order to calculate r, the correlation
coefficient, you must turn **On** the **Diagnostic** command. Press **2ⁿᵈ [CATALOG]**
(Note: CATALOG is found above the 0 key). The CATALOG of functions will
appear on the screen. Use the down arrow to scroll to the **DiagnosticOn**
command.

Press **ENTER ENTER**.

Press **STAT**, highlight **CALC**, scroll down to **4:LinReg(ax+b)** and press
ENTER ENTER. (Note: This command allows you to specify which lists
contain the X-values and Y-values. If you do not specify these lists, the defaults
are used. The defaults are: **L1** for the X-values and **L2** for the Y-values.)

```
LinReg
 y=ax+b
 a=12.48094391
 b=33.68290034
 r²=.9577738551
 r=.9786592129
```

The correlation coefficient is r = .9786592129. This indicates a strong positive
linear correlation between X and Y.

▶ Exercise 17 (pg. 508) Constructing a Scatter plot and
                        Determining r

Enter the X-values into **L1** and the Y-values into **L2**. To prepare to construct the
scatterplot, press the **Y=** key and clear all the Y-registers. Press **2ⁿᵈ [STAT
PLOT]** , select **1:Plot1**, turn **ON** Plot 1 and press **ENTER**. For **Type** of graph,
select the **scatter plot**. Enter **L1** for **Xlist** and **L2** for **Ylist**. Press **ZOOM** and **9**
to select **ZoomStat**.

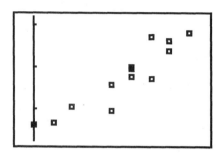

To calculate r, press **STAT**, highlight **CALC** and scroll down to
4:LinReg(ax+b) and press **ENTER ENTER**. (Note: If you have not turned **ON**
the **Diagnostics,** do that first by following the instructions for Example 5 on the
previous page.)

```
LinReg
 y=ax+b
 a=7.349665924
 b=34.6169265
 r²=.8512827753
 r=.922649866
```

The scatter plot shows a strong positive linear correlation. This is confirmed by
the r-value of 0.923.

◀

▸ Exercise 25 (pg. 510) Testing Claims

To test the significance of the population correlation coefficient, ρ, the appropriate hypothesis test is: $\rho = 0$ vs. $\rho \neq 0$. To run the test, enter the X-values into **L1** and the Y-values into **L2**. Press **STAT**, highlight **TESTS** and select **E:LinRegTTest**. Enter **L1** for **Xlist**, **L2** for **Ylist**, and **1** for **Freq**. On the next line, β and ρ, select $\neq 0$ and press **ENTER**. Leave the next line, RegEQ, blank. Highlight **Calculate.**

```
LinRegTTest
 Xlist:L₁
 Ylist:L₂
 Freq:1
 β & ρ:≠0 <0 >0
 RegEQ:
 Calculate
```

Press **ENTER**.

```
LinRegTTest
 y=a+bx
 β≠0 and ρ≠0
 t=7.935104115
 p=7.0576131ᴇ⁻6
 df=11
↓a=34.6169265
```

The output displays several pieces of information describing the relationship between X and Y. What you are interested in for this example are the following: the test statistic (t = 7.935), the P-value (p = 7.0576131E-6) and the r value (r = .922649886). Since the P-value is less than α, the correct decision is to **Reject** the null hypothesis. This indicates that there is a significant linear relationship between X and Y.

◀

Section 9.2

▶ Example 2 (pg. 515)   Finding a Regression Line

Enter the X-values into **L1** and the Y-values into **L2**. Press **STAT**, highlight
CALC and scroll down to **4:LinReg(ax+b)**. This command has several options.
One option allows you to store the regression equation into one of the Y-
variables. To use this option, with the cursor flashing on the line **LinReg(ax+b)**,
press **VARS**.

Highlight **Y-VARS**.

Select **1:Function** and press **ENTER**

Notice that **1:Y1** is highlighted. Press **ENTER**.

```
LinReg(ax+b) Y1
```

Press **ENTER**.

```
LinReg
  y=ax+b
  a=12.48094391
  b=33.68290034
  r²=.9577738551
  r=.9786592129
```

The output displays the general form of the regression equation: y = ax+b
followed by values for a and b. Next, r^2, the coefficient of determination, and r,
the correlation coefficient , are displayed. If you put the values of a and b into
the general equation, you obtain the specific linear equation for this data:
y= 12.48 x + 33.68. Press **Y=** and see that this specific equation has been pasted
to **Y1**.

```
Plot1  Plot2  Plot3
\Y1☐12.480943910
438X+33.68290034
0426
\Y2=
\Y3=
\Y4=
\Y5=
```

Press **2ⁿᵈ** **[STAT PLOT]** , select **1:Plot1**, turn **ON** Plot1, select **scatter plot**, set
Xlist to **L1** and **Ylist** to **L2**. Press **ZOOM** and **9**.

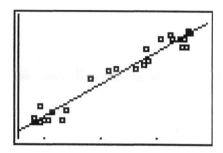

This picture displays a scatter plot of the data and the regression line. The picture indicates a strong positive linear correlation between X and Y, which is confirmed by the r-value of .979.

You can use the regression equation stored in **Y1** to predict Y-values for specific X-values. For example, suppose the duration of an eruption was equal to 1.95 minutes. Predict the time (in minutes) until the next eruption. In other words, for X = 1.95, what does the regression equation predict for Y? To find this value for Y, press **VARS**, highlight **Y-VARS**, select **1:Function**, press **ENTER**, select **1:Y1** and press **ENTER**. Press **(** 1.95 **)** and press **ENTER** .

```
Y₁(1.95)
        58.02074097
```

The output is a display of the predicted Y-value for X = 1.95.

▶ Exercise 13 (pg. 518) Finding the Equation of the Regression Line

Enter the X-values into **L1** and the Y-values into **L2**. Press **STAT**. Highlight **CALC**, select **4:LinReg(ax+b)**, press **ENTER**. Press **VARS**, highlight **Y-VARS**, select **1:Function**, press **ENTER** and select **1:Y1** and press **ENTER**.

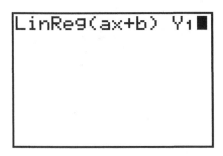

```
LinReg(ax+b) Y1█
```

Press **ENTER**.

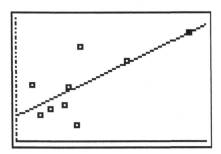

```
LinReg
 y=ax+b
 a=.0619401188
 b=7.534504806
 r²=.5320119719
 r=.729391508
█
```

Using **a** and **b** from the output display, the resulting regression equation is $y = 0.06194 x + 7.5345$. Press **Y=** to confirm that the regression equation has been stored in **Y1**.

To view the data and the regression line, first make sure that the scatterplot has been selected. Press **2ⁿᵈ** **[STAT PLOT]** , select **1:Plot1**, turn **ON** Plot1, select **scatter plot**, set **Xlist** to **L1** and **Ylist** to **L2**. Press **ZOOM** and **9** and a graph of the scatter plot with the regression line will be displayed.

Next, you can use the regression equation to predict number of stories, Y, for various heights, X. First, check the X-values that you will be using to confirm that they are within (or close to) the range of the X-values in your data. Three X-values (500, 650 and 725) meet this criteria. The value of 310 is outside the range of the original values, so the regression equation will not be used with this value.

Press **VARS**, highlight **Y-VARS**, select **1:Function** and press **ENTER**. Select **1:Y1** and press **ENTER**. Press **(** 500 **)** and **ENTER**.

```
Y₁(500)
         38.50456422




```

The predicted number of stories for a building with height of 500 ft. is 39 stories.

Press **2ⁿᵈ** **[ENTRY]**, (found above the ENTER key). Move the cursor so that it is flashing on '**5**' in the number '500' and type in 650. Press **ENTER**.

```
Y₁(500)
         38.50456422
Y₁(650)
         47.79558205


```

The predicted number of stories for a building with a height of 650 ft. is 48 stories.

Press **2ⁿᵈ** **[ENTRY]**. Move the cursor so that it is flashing on '**6**' in the number '650' and type in 725. Press **ENTER**.

The predicted number of stories for a building with a height of 725 ft. is 52 stories.

Section 9.3

▶ Example 2 (pg. 528)   The Standard Error of the Estimate

Enter the data into **L1** and **L2**. Press **STAT**, highlight **CALC**, select **4:LinReg(ax+b)**, and press **ENTER**. Press **VARS**, highlight **Y-VARS**, select **1:Function**, press **ENTER**, select **1:Y1** and press **ENTER** **ENTER**.

The formula for s_e, the standard error of the estimate is $\sqrt{\dfrac{\sum(y_i - \hat{y}_i)^2}{n-2}}$. The values for $(y_i - \hat{y}_i)$, called Residuals, are automatically stored to a list called **RESID**. Press **2ⁿᵈ** **[LIST]**, select **RESID**. Press **STO** **2ⁿᵈ** **[L3]** and **ENTER**. This stores the residuals to **L3**.

```
LRESID→L₃
{-.8097165992  -…
```

In the formula for s_e, the residuals, $(y_i - \hat{y}_i)$, are squared. To square these values and store them in **L4**, press **STAT**, select **1:Edit** and move the cursor to highlight the Listname **L4**. Press **ENTER**. Press **2ⁿᵈ** **[L3]** and the x^2 key.

L2	L3	**L4**	4
225	-.8097	------	
184	-1.227		
220	14.482		
240	4.0445		
180	4.919		
184	-1.227		
186	-19.52		

$L4 = L_3^2$

Press **ENTER**.

```
L2        L3        L4        4
225       -.8097    .65564
184       -1.227    1.5048
220       14.482    209.72
240       4.0445    16.358
180       4.919     24.197
184       -1.227    1.5048
186       -19.52    380.96
L4(1)=.6556409710...
```

Press 2nd **[QUIT]** . Next, you need to find the sum of **L4**, $\sum (y_i - \hat{y}_i)^2$. Press 2nd **[LIST]** . Highlight **MATH**, select **5:sum(** and press 2nd **[L4]**. Close the parentheses and press **ENTER**.

```
sum(L4)
            635.3441296
```

Divide this sum by (number of observations – 2). In this example, (n - 2) = 6. Simply press ÷ 6 and press **ENTER**.

```
sum(L4)
            635.3441296
Ans/6
            105.8906883
```

Lastly, take the square root of this answer by pressing 2nd $[\sqrt{\ }]$ 2nd **[ANS]**. Close the parentheses and press **ENTER**.

```
sum(L4)
          635.3441296
Ans/6
          105.8906883
√(Ans)
          10.29032012
■
```

The standard error of the estimate, s_e, is 10.29.

◄

▶ Example 3 (pg. 530) Constructing a Prediction Interval

This example is a continuation of Example 2 on pg. 528. To construct a 95%
prediction interval for a specific X-value, x_o, you must calculate the maximum

error, E. The formula for E is: $E = t_c s_e \sqrt{(1 + \dfrac{1}{n} + \dfrac{n(x_o - \bar{x})^2}{n(\sum x^2) - (\sum x)^2}}$. The

critical value for t is found using the inverse-t Distribution function on the TI-84
calculator. (Note: If you are using a TI-83 calculator, you cannot find the t-value
on the calculator. Use a t-table to find the t-value.) On the TI-84, press **2ⁿᵈ**
[DISTR] and select **4:invT.** Type in the cumulative area for the upper end of the
95 % confidence interval: .975, and the degrees of freedom: n – 1 = 6. Press
ENTER.

```
invT(.975,6)
         2.446911839
```

For this example, t_c = 2.447. The standard error, s_e, is 10.290. (Note: The
standard error was calculated in Example 2 on the previous pages.)

To calculate $\bar{x}$, $\sum x^2$, and $(\sum x)^2$, press **VARS**, select **5:Statistics**.
Highlight **2:$\bar{x}$** and press **ENTER ENTER**. Notice that $\bar{x}$ = 1.975. Press **VARS**
again, select **5:Statistics**, highlight $\sum$, select **1:$\sum x$** and press **ENTER**
ENTER. So, $\sum x$ = 15.8. Press **VARS** again, select **5:Statistics**, highlight $\sum$,
select **2:$\sum x^2$** , and press **ENTER ENTER**. Notice that $\sum x^2$ = 32.44.

Next, calculate the maximum error, E, when x_o = 2.1 using the formula for E.

```
2.447*10.29*√(1+
1/8+8(2.1-1.975)
²/(8*32.44-15.8²
)
        26.85678531
■
```

The prediction interval is $\hat{y} \pm 26.857$. To find $\hat{y}$, the predicted value for y when x = 2.1, press **VARS**, highlight **Y-VARS**, select **1:Function** and press **ENTER**. Next select **1:Y1**, press **ENTER** and press **(** 2.1 **)**. Finally, press **ENTER**.

```
2.447*10.29*√(1+
1/8+8(2.1-1.975)
²/(8*32.44-15.8²
)
        26.85678531
Y₁(2.1)
        210.5910931
■
```

The prediction interval is 210.59 ± 26.857.

> ▶ Exercise 11 (pg. 532)   Coefficient of Determination and
>   Standard Error of the Estimate

Enter the data into **L1** and **L2**. Press **STAT**, highlight **CALC**, select
4:LinReg(ax+b) and press **ENTER**. Next, press **VARS**, select **Y-VARS**, select
1:Function and press **ENTER**. Lastly, select **1:Y1** and press **ENTER ENTER**.

```
LinReg
 y=ax+b
 a=549.4482536
 b=-1881.69392
 r²=.9805552664
 r=.9902299058
■
```

The coefficient of determination, r^2, is .9806. This means that 98.06 % of the
variationin the Y-values is explained by the X-values.

To calculate s_e, press **2ⁿᵈ** **[LIST]** , select **RESID**, press **ENTER**, **STO** , **2ⁿᵈ**
[L3] **ENTER**. This stores the residuals to **L3**. In the formula for s_e, the
residuals, $(y_i - \hat{y}_i)$, are squared. To square these values and store them in **L4**,
press **STAT**, select **1:Edit** and move the cursor to highlight the Listname **L4**.
Press **ENTER**. Press **2ⁿᵈ** **[L3]** and the x^2 key. Press **ENTER**.

```
L2       L3       L4        4
893.8    28.253   798.34
933.9    13.408   179.77
980      4.563    20.821
1032.4   2.0182   4.073
1105.3   -34.97   1223
1181.1   -14.12   199.27
1221.7   -28.46   810.04
L4(1)=798.2123453...
```

Press **2ⁿᵈ** **[QUIT]** . Next, you need to find the sum of **L4**, $\sum (y_i - \hat{y}_i)^2$. Press
2ⁿᵈ **[LIST]** . Highlight **MATH**, select **5:sum(** and press **2ⁿᵈ** **[L4]**. Close the
parentheses and press **ENTER**.

```
  r²=.9805552664
  r=.9902299058

∟RESID→L₃
{28.25265201 13…
sum(L₄)
        8413.9576
■
```

Divide this sum by (number of observations – 2). In this example, (n - 2) = 9. Simply press ÷ 9 and press ENTER.

```
∟RESID→L₃
{28.25265201 13…
sum(L₄)
        8413.9576
Ans/9
      934.8841778
■
```

Lastly, take the square root of this answer by pressing **2ⁿᵈ** [√] **2ⁿᵈ** **[ANS]**. Close the parentheses and press ENTER.

```
{28.25265201 13…
sum(L₄)
        8413.9576
Ans/9
      934.8841778
√(Ans)
      30.57587575
```

The standard error of the estimate is $30.576 billion.

▸ Technology (pg. 549)   Nutrients in Breakfast Cereals

Exercises 1-2: Enter the data into **L1, L2, L3** and **L4**. To prepare to construct the scatterplots, press the **Y=** key and clear all the Y-registers. To construct the scatter plots, press **2ⁿᵈ [STAT PLOT]**, select **1:Plot1** and press **ENTER**. Turn **ON** Plot1. Select **scatter plot** for **Type**. Enter the appropriate labels for **Xlist** and **Ylist** to construct each of the scatter plots.

Exercises 3: To find the correlation coefficients, press **STAT**, highlight **CALC** and select **4:LinReg(ax+b)** and enter the labels of the columns you are using for the correlation. For example, to find the correlation coefficient for "fat" and "carbohydrates", press **2ⁿᵈ [L3] , 2ⁿᵈ [L4]**.

```
LinReg(ax+b) L3,
L4█
```

Press **ENTER** and notice that the correlation coefficient for fat and carbohydrates is -.021.

Exercise 4: To find the regression equations, press **STAT**, highlight **CALC** and select **4:LinReg(ax+b)** and enter the labels of the columns you are using for the regression. For example, to find the regression equation for "calories" and "carbohydrates", press **2ⁿᵈ [L1] , 2ⁿᵈ [L4]** .

Exercise 5: Use the regression equations found in Exercise 4 to do the predictions.

Exercises 6 and 7: The TI-83/84 does not do multiple regression.

◀

Getting Started with Microsoft Excel

Beverly Dretzke
University of Minnesota

FOURTH EDITION

Elementary Statistics
Picturing *the* World

Larson | Farber

PEARSON

Prentice
Hall

Upper Saddle River, NJ 07458

▶ Contents:

Getting Started with Microsoft Excel 2007

Overview

This manual is intended as a companion to Larson and Farber's *Elementary Statistics, 4th ed.* It presents instructions on how to use Microsoft Excel 2007 to carry out selected examples and exercises from *Elementary Statistics, 4th ed.*

The first section of the manual contains an introduction to Microsoft Excel 2007 and how to perform basic operations such as entering data, using formulas, saving worksheets, retrieving worksheets, and printing. All the screens pictured in this manual were obtained using the Office 2007 version of Microsoft Excel on a PC. You may notice slight differences if you are using a different version or a different computer.

Getting Started with the User Interface

GS 1.1 The Mouse

The mouse is a pointer device that allows you to move around the Excel worksheet and to select specific locations and objects. There are four main mouse operations: Select, click, double-click, and right-click.

1. To **select** generally means to move the mouse pointer so that the white arrow is pointing at or is positioned directly over an object. You will often **select** commands in the Office icon menu. Some of the more familiar of these commands are Open, Save, and Print.

2. To **click** means to press down on the left button of the mouse. You will frequently select cells of the worksheet and commands by "clicking" the left button.

3. To **double-click** means to press the left mouse button twice in rapid succession.

4. To **right-click** means to press down on the right button of the mouse. A right-click is often used to display special shortcut menus.

GS 1.2	The Excel 2007 Window

The figure shown below presents the top left side of the Exel 2007 window. The Office icon is located in the upper-left corner. Near the top of the screen, you see a row of several tabs: Home, Insert, Page Layout, Formulas, Data, Review, View, and Add-Ins. Each tab leads to a ribbon. The Home ribbon, shown below, presents groups of related commands. The groups that are displayed in the figure are Clipboard, Font, Alignment, and Number.

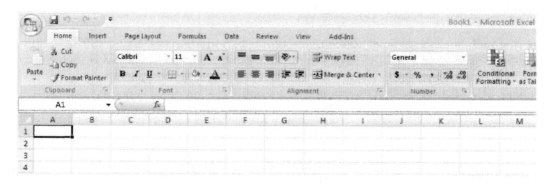

Clipboard commands may be replaced by keyboard shortcuts for those of you who prefer to use them. Some of the more familiar of these shortcuts are Ctrl+X for cut, Ctrl+C for copy, and Ctrl+V for paste.

GS 1.3	Ribbons and the Office Icon

When you click one of the tabs at the top of the screen, you will see a ribbon of related commands. The **Home ribbon** is displayed above in GS 1.2.

The **Insert** ribbon, shown below, includes groups of commands for Tables, Illustrations, Charts, Links, and Text.

The **Page Layout ribbon** includes groups of commands for Themes, Page Setup, Scale to Fit, Sheet Options, and Arrange.

The **Formulas ribbon** includes groups of commands for Function Library, Defined Names, Formula Auditing, and Calculation.

The **Data ribbon** includes groups of commands for Get External Data, Connections, Sort & Filter, Data Tools, Outline, and Analysis.

The **Office icon** in the top left corner of the screen contains the commands New, Open, Save, Save As, Print, etc. Click the Office icon to display the menu.

Office icon

GS 1.4 | Dialog Boxes

Many of the statistical analysis procedures that are presented in this manual are associated with commands that are followed by dialog boxes. Dialog boxes usually require that you select from alternatives that are presented or that you enter your choices.

For example, if you click the **Formulas** tab and select **Insert Function**, a dialog box like the one shown below will appear. You make your function category and function name selections by clicking on them. The SLOPE function that has been selected in this example is found in the Statistical category. If you want to find out what functions are available, you can type in a brief description of what you would like to do and then click the Go button.

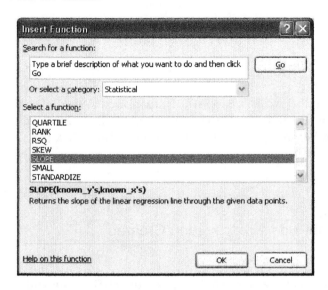

When you click the OK button at the bottom of the dialog box, another dialog box will often be displayed that asks you to provide information regarding location of the data in the Excel worksheet

Getting Started with Opening Files

GS 2.1	Opening a New Workbook

When you start Excel, the screen will open to **Sheet 1** of **Book 1**. Sheet names appear on tabs at the bottom of the screen. The name "Book 1" will appear at the top.

If you are already working in Excel and have finished the analyses for one problem and would like to open a new book for another problem, follow these steps: First, at the top of the screen, click the **Office icon** and select **New**.

Next, double-click **Blank Workbook** (or click **Create** in the bottom right corner). If you were previously working in Book1, the new worksheet will be given the default name Book 2.

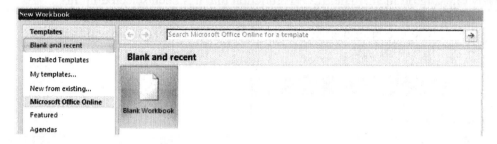

The names of books opened during an Excel work session will be displayed in the Window group of the View ribbon. To return to one of these books, click the **View** tab at the top of the screen and select **Switch Windows** in the Window group. Then click the book name.

GS 2.2	Opening a File That Has Already Been Created

To open a file that you or someone else has already created, click on the **Office icon** and select **Open**. A list of file locations will appear. Select the location by clicking on it. Many of the data files that are presented in your statistics textbook are available on the CD that accompanies this manual. To open any of these files, you will select the CD drive on your computer.

After you click on the CD drive, a list of folders and files available on the CD will appear. You will need to select the folder or file you want by clicking on it. If you have selected a folder, another screen will appear with a list of files contained in the folder. Click on the name of the file that you would like to open.

Getting Started with Entering Information

GS 3.1	Cell Addresses

Columns of the worksheet are identified by letters of the alphabet and rows are identified by numbers. The cell address A1 refers to the cell located in column A row 1. The dark outline around a cell means that it is "active" and is ready to receive information. In the figure shown below, cell C1 is ready to receive information. You can also see C1 in the **Name Box** to the left of the **Formula Bar**. You can move to different cells of the worksheet by using the mouse pointer and clicking on a cell. You can also press [**Tab**] to move to the right or left, or you can use the arrow keys on the keyboard.

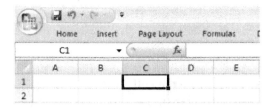

You can also activate a **range** of cells. To activate a range of cells, first click in the top cell and drag down and across (or click in the bottom cell and drag up and across). The range of cells highlighted in the figure below is designated B2:D4.

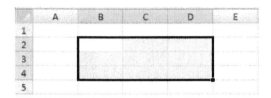

| GS 3.2 | Types of Information |

Three types of information may be entered into an Excel worksheet.

1. **Text**. The term "text" refers to alphabetic characters or a combination of alphabetic characters and numbers, sometimes called "alphanumeric." The figure provides an example of an entry comprised solely of alphabetic characters (cell A1) and an entry comprised of a combination of alphabetic characters and numbers (cell B1).

	A	B
1	Sue Clark	25 years
2		

2. **Numeric**. Any cell entry comprised completely of numbers falls into the "numeric" category.

	A
1	2
2	9
3	8
4	5

3. **Formulas**. Formulas are a convenient way to perform mathematical operations on numbers already entered into the worksheet. Specific instructions are provided in this manual for problems that require the use of formulas.

GS 3.3	Entering Information

To enter information into a cell of the worksheet, first activate the cell. Then key in the desired information and press [**Enter**]. Pressing the [Enter] key moves you down to the next cell in that column. The information shown below was entered as follows:

1.   Click in cell A1. Key in **1**. Press [**Enter**].

2.   Key in **2**. Press [**Enter**].

3.   Key in **3**. Press [**Enter**].

	A	B
1	1	
2	2	
3	3	
4		
5		

GS 3.4	Using Formulas

When you want to enter a formula, begin the cell entry with an equal sign (=). The arithmetic operators are displayed below.

Arithmetic operator	Meaning	Example
+	Addition	=3+2
-	Subtraction	=3-2
*	Multiplication	=3*2
/	Division	=3/2
^	Exponentiation	=3^2

Numbers, cell addresses, and functions can be used in formulas. For example, to sum the contents of cells A1 and B1, you can use the formula =A1+B1. To divide this sum by 2, you can use the formula =(A1+B1)/2. Note that Excel carries out expressions in parentheses first and then uses the results to complete the calculations of the formula. Formulas will sometimes not produce the desired results because parentheses were necessary but were not used.

Getting Started with Changing Information

GS 4.1	Editing Information in the Cells

There are several ways that you can edit information that has already been entered into a cell.

1. If you have not completed the entry, you can simply backspace and start over. Clicking on the red X to the left of the Formula Bar will also delete an incomplete cell entry.

2. If you have already completed the entry and another cell is activated, you can click in the cell you want to edit and then press either [**Delete**] or [**Backspace**] to clear the contents of the cell.

3. If you want to edit part of the information in a cell instead of deleting all of it, follow the instructions provided in the example.

    •   Let's say that you wanted to enter 1234 in cell A1 but instead entered 124. Return to cell **A1** to make it the active cell by either clicking on it with the mouse or by using the arrow keys.

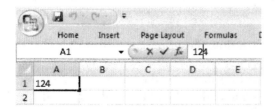

    •   You will see A1 in the Name Box and 124 in the Formula Bar. Click between 2 and 4 in the Formula Bar so that the **I-beam** is positioned there. Enter the number 3 and press [**Enter**].

GS 4.2	Copying Information

To copy the information in one cell to another cell, follow these steps:

- First click on the source cell. Then, at the top of the screen, click the **Home** tab and select **Copy**.

- Click on the target cell where you want the information to be placed. Then, at the top of the screen, click **Home** and select **Paste**.

To copy a range of cells to another location in the worksheet, follow these steps:

- First click and drag over the range of cells that you want to copy so that they are highlighted. Then, at the top of the screen, click **Home** and select **Copy**.

- Click in the topmost cell of the target location. Then, at the top of the screen, click **Home** and select **Paste**.

To copy the contents of one cell to a range of cells follow these steps:

- Let's say that you have entered a formula in cell C1 that adds the contents of cells A1 and B1 and you would like to copy this formula to cells C2 and C3 so that C2 will contain the sum of A2 and B2 and cell C3 will contain the sum of A3 and B3.

- First click in cell C1 to make it the active cell. You will see =A1+B1 in the Formula Bar.

C1			f_x	=A1+B1	
	A	B	C	D	E
1	1	1	2		
2	2	2			
3	3	3			

- At the top of the screen, click **Home** and select **Copy**.

- Highlight cells C2 and C3 by clicking and dragging over them.

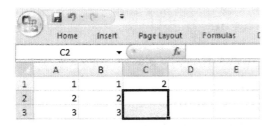

- At the top of the screen, click **Home** and select **Paste**. The sums should now be displayed in cells C2 and C3.

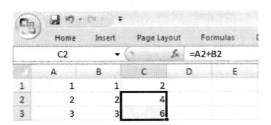

GS 4.3	Moving Information

If you would like to move the contents of one cell from one location to another in the worksheet, follow these steps:

- Click on the cell containing the information that you would like to move.

- At the top of the screen, click **Home** and select **Cut**.

- Click on the target cell where you want the information to be placed.

- At the top of the screen, click **Home** and select **Paste**.

If you would like to move the contents of a range of cells to a different location in the worksheet, follow these steps:

- Click and drag over the range of cells that you would like to move so that it is highlighted.

- At the top of the screen, click **Home** and select **Cut**.

- Click the topmost cell of the new location. (It is not necessary to click and drag over the entire range of the new location.)

- At the top of the screen, click **Home** and select **Paste**.

*If you make a mistake, just click the **Undo** arrow located in the upper-left corner of the screen. It looks like this:*

GS 4.4	Changing the Column Width

There are a couple of different ways that you can use to change the column width. Only one way will be described here. Output from the Descriptive Statistics data analysis tool will be used as an example. As you can see in the output displayed below, many of the labels in column A can only be partially viewed because the default column width is too narrow.

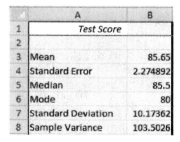

	A	B
1	*Test Score*	
2		
3	Mean	85.65
4	Standard I	2.274892
5	Median	85.5
6	Mode	80
7	Standard I	10.17362
8	Sample V₂	103.5026

Position the mouse pointer directly on the vertical line between A and B in the letter row at the top of the worksheet — A | B —so that it turns into a black plus sign.

Click and drag to the right until you can read all the output labels. (You can also click and drag to the left to make columns narrower.) After adjusting the column width, your output should appear similar to the output shown below.

	A	B
1	*Test Score*	
2		
3	Mean	85.65
4	Standard Error	2.274892
5	Median	85.5
6	Mode	80
7	Standard Deviation	10.17362
8	Sample Variance	103.5026

Getting Started with Sorting Information

GS 5.1   Sorting a Single Column of Information

Let's say that you have entered "Score" in cell A1 and four numbers directly below it and that you would like to sort the numbers in ascending order.

	A
1	Score
2	15
3	79
4	18
5	2

- Click and drag from cell A1 to cell A5 so that the range of cells is highlighted.

You could also click directly on 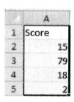 *in the letter row at the top of the worksheet. This will result in all cells of column A being highlighted.*

	A
1	Score
2	15
3	79
4	18
5	2

- At the top of the screen, click the **Data** tab and select **Sort** in the Sort & Filter Group.

- In the Sort dialog box that appears, you are given the choice of sorting the information in column A in either Smallest to Largest or Largest to Smallest order. The **Smallest to Largest** order has already been selected. There is a checkmark in the box to the left of **My data has headers**. This means that the "Score" header will stay in cell A1 and will not be included in the sort. Click **OK** at the bottom of the dialog box.

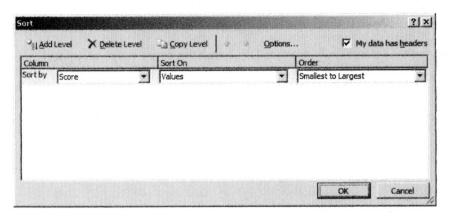

The cells in column A should now be sorted in ascending order as shown below.

	GS 5.2	Sorting Multiple Columns of Information

Your Excel data files will frequently contain multiple columns of information. When you sort multiple columns at the same time, Excel provides a number of options.

Let's say that you have a data file that contains the information shown below and that you would like to sort the file by GPA in descending order.

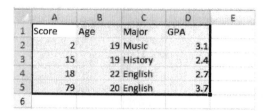

	A	B	C	D
1	Score	Age	Major	GPA
2	2	19	Music	3.1
3	15	19	History	2.4
4	18	22	English	2.7
5	79	20	English	3.7

- Click and drag from A1 down and across to D5 so that the entire range of cells is highlighted.

	A	B	C	D	E
1	Score	Age	Major	GPA	
2	2	19	Music	3.1	
3	15	19	History	2.4	
4	18	22	English	2.7	
5	79	20	English	3.7	
6					

- At the top of the screen, click the **Data** tab at the top of the screen and select **Sort** in the Sort & Filter group.

- In the Sort dialog box that appears, you are given the option of sorting the data by four different variables. You want to sort only by GPA in descending order. Click the down arrow to the right of the Sort by window until you see GPA and click on **GPA** to select it. Select the **Largest to Smallest** order. The checkmark in the box to the left of **My data has headers** means that the top row in the selected range will not be included in the sort, and the variable labels will stay in row 1. Click **OK**.

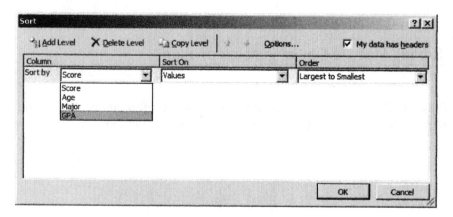

The sorted data file is shown below.

	A	B	C	D
1	Score	Age	Major	GPA
2	79	20	English	3.7
3	2	19	Music	3.1
4	18	22	English	2.7
5	15	19	History	2.4

Getting Started with Saving Information

GS 6.1	Saving Files

To save a newly created file for the first time, click the **Office icon** in the upper-left corner of the screen and select **Save**. A Save As dialog box will appear. You will need to select the location for saving the file by clicking on it.

The default file name, displayed in the File name window, is **Book1.** It is highly recommended you replace the default name with a name that is more descriptive.

Once you have saved a file, clicking the **Office icon** and **Save** will result in the file being saved in the same location under the same file name. No dialog box will appear. If you would like to save the file in a different location, you will need to select click the **Office icon** and select **Save As**.

GS 6.2	Naming Files

Windows and Mac versions of Excel will allow file names to have around 200 characters. You will find that long, descriptive names will be easier to work with than really short names. For example, if a file contains data that was collected in a survey of Milwaukee residents, you may want to name the file **Milwaukee resident survey** rather than **MRS.**

Several symbols cannot be used in file names. These include: forward slash (/), backslash (\), greater-than sign (>), less-than sign (<), asterisk (*), question mark (?), quotation mark ("), pipe symbol (|), color (:), and semicolon (;).

Getting Started with Printing Information

GS 7.1   ## Printing Worksheets

To print a worksheet, click the **Office icon** in the upper-left corner of the screen and select **Print**. The Print dialog box will appear.

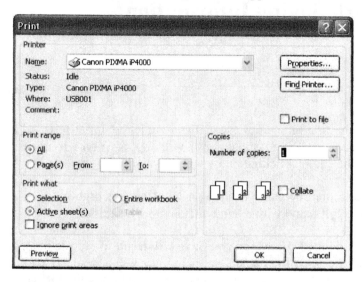

Under Print range, you will usually select **All**, and under Print what, you will usually select **Active sheet(s)**. The default number of copies is 1, but you can increase this if you need more copies. When the Print dialog box has been completed as you would like, click **OK**.

GS 7.2   ## Page Setup

Excel provides a number of page setup options for printing worksheets. To access these options, click the **Page Layout** tab at the top of the screen and select from the command groups. In **Page Setup**, you may want to select **Orientation** to change from portrait to landscape orientation for worksheets that have several columns of data. Under the **Gridlines** command in the **Sheet Options** groups, you may want to select **Print** so that gridlines appear in your hard copies.

Getting Started with Add-ins

 GS 8.1   Loading Excel's Analysis Toolpak

The Analysis Toolpak is an Excel Add-In that may not necessarily be loaded. If it does not appear in the Analysis group of the Data ribbon as shown below, then you will need to load it.

First click the **Office icon** and, at the bottom of the dialog box, select **Excel Options**. Select **Add-Ins** from the list on the left. Analysis ToolPak and Analysis ToolPak – VBA should both be in the list of Active Application Add-Ins. If you need to add one or both of them, first click on the name to select it. Then click **Go** at the bottom of the screen. In the Add-Ins dialog box, place a checkmark in the box next to the add-in to make it active. Click **OK**.

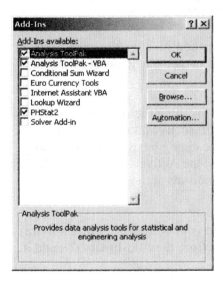

| GS 8.2 | Loading the PHStat2 Add-In |

PHStat2 is a Prentice Hall statistical add-in that is included on the CD-ROM that accompanies this manual. The instructions that are given here also appear in the PHStat readme file.

To use the Prentice Hall PHStat2 Microsoft Excel add-in, you first need to run the setup program (Setup.exe) located in the PHStat2 folder on the CD-ROM. The setup program will install the PHStat2 program files to your system and add icons on your Desktop and Start Menu for PHStat2. Depending on the age of your Windows system files, some Windows system files may be updated during the setup process as well. Note that PHStat2, Version 2.7.0, is compatible with Microsoft Excel 2000, 2002(XP), 2003, and 2007. PHStat2, Version 2.7.0, is not compatible with Microsoft Excel 95 or 97.

During the Setup program you will have the opportunity to specify:

a. The directory into which to copy the PHStat2 files (default is \Program Files\Prentice Hall\PHStat2).

b. The name of the Start Programs folder to contain the PHStat2 icon.

To begin using the Prentice Hall PHStat2 Microsoft Excel add-in, click the appropriate Start Menu or Desktop icon for PHStat2 that was added to your system during the setup process.

When a new, blank Excel worksheet appears, make sure that both the **Analysis ToolPak** and **Analysis ToolPak–VBA** add-ins are active. To see the active add-ins, click the **Office icon** and select **Excel Options** at the bottom of the dialog box. Select **Add-Ins** from the list on the left. A list of add-ins will appear, and the Active Application Add-Ins will be shown at the top. If you need to make these two add-ins active, see the instructions in section GS 8.1 of this manual.

You can start PHStat2 directly from the program menu in your computer. I find it more convenient to make PHStat2 an add-in and then access it from the Add-In tab. To add it, follow these steps.

1. Click the **Office icon** and select **Excel Options**.

2. Select **Add-Ins** from the list on the left.

3. Near the bottom, you will see Manage Excel Add-Ins. Click **Go.**

4. Click **Browse** in the Add-Ins dialog box and locate the executable PHStat2 file on your computer. It is most likely in the Program Files folder. Select it and click **OK.**

5. Click in the box to the left of PHStat2 to make it an active add-in. Click **OK.**

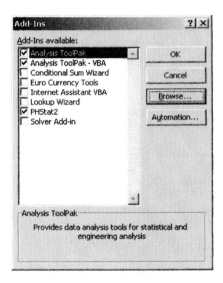

6. When you click the Add-Ins tab at the top of your screen, you will see PHStat2 in the Menu Commands group.

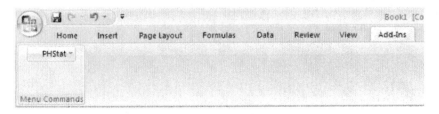

7. When you click **PHStat**, you will see the PHStat2 menu choices.

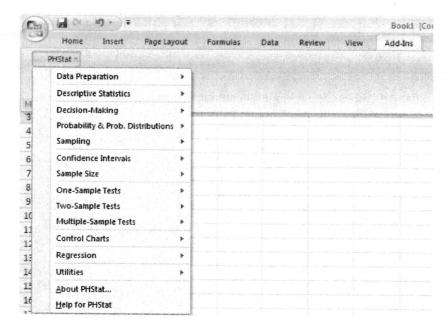

Introduction to Statistics

Technology

▶ Example (pg. 36)	Generating a List of Random Numbers

You will be generating a list of 15 random numbers between 1 and 167 to use in selecting a random sample of 15 cars assembled at an auto plant.

If the PHStat2 add-in has not been loaded, you will need to load it before continuing. Follow the instructions in Section GS 8.2.

---

1. Open a new Excel worksheet.

2. At the top of the screen, click **Add-Ins** and select **PHStat**. Select **Sampling→Random Sample Generation**.

3. Complete the Random Sample Generation dialog box as shown below. A sample of 15 cars will be randomly selected from a population of 167. The topmost cell in the output will contain the title "Car #." Click **OK**.

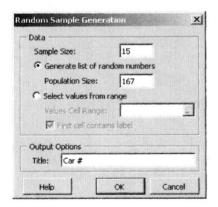

The output is displayed in a new worksheet named "RandomNumbers." Because the numbers were generated randomly, it is not likely that your output will be exactly the same.

	A
1	Car #
2	144
3	139
4	63
5	129
6	25
7	39
8	94
9	32
10	14
11	26
12	140
13	44
14	159
15	98
16	47

◄

► Exercise 1 (pg. 37)	Generating a List of Random Numbers for the SEC

You will be generating a list of 10 random numbers between 1 and 86 to use in selecting a random sample of brokers. You will also be ordering the generated list from lowest to highest.

If the PHStat2 add-in has not been loaded, you will need to load it before continuing. Follow the instructions in Section GS 8.2.

1. Open a new Excel worksheet.

2. At the top of the screen, click **Add-Ins** and select **PHStat**. Select **Sampling→Random Sample Generation**.

3. Complete the Random Sample Generation dialog box as shown below. A sample of 10 accounts will be randomly selected from a population of 86. "Broker #" will appear in the top cell of the output. Click **OK**.

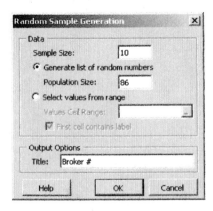

4. Sort the numbers in ascending order.

For instructions on how to sort, refer to Sections GS 5.1 and GS 5.2.

	A
1	Broker #
2	30
3	61
4	59
5	76
6	40
7	7
8	43
9	83
10	4
11	50

The sorted set of ten random numbers is displayed below. Because the numbers were generated randomly, it is not likely that your output will be exactly the same.

	A
1	Broker #
2	4
3	7
4	30
5	40
6	43
7	50
8	59
9	61
10	76
11	83

◄

► Exercise 2 (pg. 37)	Generating a List of Random Numbers for a Quality Control Department

If the PHStat2 add-in has not been loaded, you will need to load it before continuing. Follow the instructions in Section GS 8.2.

1. Open a new Excel worksheet.

2. At the top of the screen, click **Add-Ins** and select **PHStat**. Select **Sampling→Random Sample Generation**.

3. Complete the Random Sample Generation dialog box as shown at the top of the next page. A sample of 25 camera phones will be randomly selected from a population of 300. "Camera Phone #" will appear in the top cell of the generated output. Click **OK**.

4. Sort the numbers in ascending order.

For instructions on how to sort, see Sections GS 5.1 and GS 5.2.

The sorted set of 25 random numbers is displayed below. Because the numbers were generated randomly, it is unlikely that your output will be exactly the same.

	A	B
1	Camera Phone #	
2	19	
3	75	
4	26	
5	239	
6	197	
7	287	
8	170	
9	178	
10	202	
11	252	
12	69	
13	253	
14	111	
15	145	
16	288	
17	130	
18	227	
19	16	
20	171	
21	121	
22	68	
23	289	
24	222	
25	247	
26	271	

| ▶ Exercise 3 (pg. 37) | Generating Three Random Samples from a Population of Ten Digits |

You will be generating three random samples of five digits each from the population: 0, 1, 2, 3, 4, 5, 6, 7, 8, 9. You will also compute the average of each sample.

If the PHStat2 add-in has not been loaded, you will need to load it before continuing. Follow the instructions in Section GS 8.2.

---

1. Open a new Excel worksheet. Enter the numbers 0 through 9 in column A.

	A
1	0
2	1
3	2
4	3
5	4
6	5
7	6
8	7
9	8
10	9

2. Compute the average using the AVERAGE function. To do this, first click in the cell immediately below the last number in the population, cell **A11**, where you will place the average. Then, at the top of the screen click **Formulas** and select **Insert Function**.

3. Select the **Statistical** category. Select the **AVERAGE** function. Click **OK**.

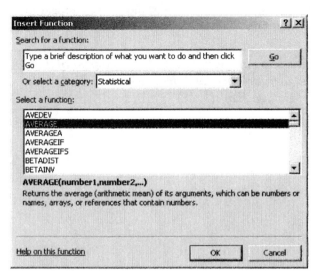

4. Complete the AVERAGE dialog box as shown below. A1:A10 is the worksheet location of the population of ten digits. Click **OK**.

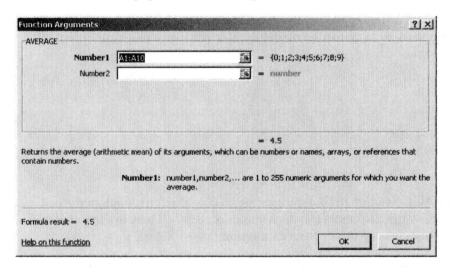

The average, 4.5, is displayed in cell A11 of the worksheet.

9	8
10	9
11	4.5

5. You will now select the first random sample of five digits. At the top of the screen, click **Add-Ins** and select **PHStat**. Select **Sampling→Random Sample Generation**.

6. Complete the Random Sample Generation dialog box as shown below. Five numbers will be randomly selected from the numbers displayed in cells A1 through A10 in the worksheet. Cell A1 does not contain a label. "Sample 1" will appear in the top cell of the generated output. Click **OK**.

A quick way to enter the range is to first click in the Values Cell Range field of the dialog box and then click on cell A1 in the worksheet and drag down to cell A10. Be careful that you do not include cell A11 in the range.

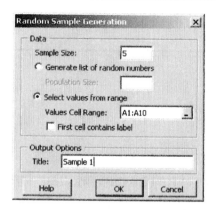

The random sample of five digits is placed in a sheet named "RandomNumbers." Because the numbers were generated randomly, it is not likely that your output will be exactly the same.

	A
1	Sample 1
2	0
3	8
4	9
5	2
6	4

7. Use the AVERAGE function to find the average. To do this, first click in the cell immediately below the last number in the sample—cell **A7**. Then, at the top of the screen, click **Formulas** and select **Insert Function**.

8. Select the **Statistical** category. Select the **AVERAGE** function. Click **OK**.

9. Complete AVERAGE dialog box as shown at the top of the next page. Click **OK**.

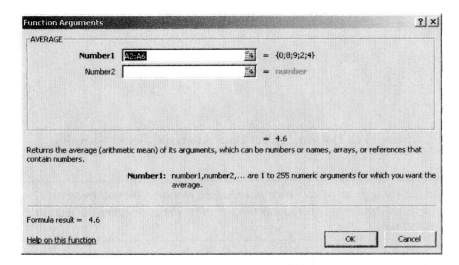

10. The average of the sample, 4.6, is now displayed in cell A7 of the worksheet. Repeat steps 5-9 to generate two more samples and compute the averages.

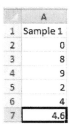

	A
1	Sample 1
2	0
3	8
4	9
5	2
6	4
7	4.6

To return to the sheet that contains the population values, click on the Sheet1 tab located near the bottom of your screen.

► Exercise 5 (pg. 37) | Simulating Rolling a Six-Sided Die 60 Times

You will be simulating rolling a six-sided die 60 times and making a tally of the results.

If the PHStat2 add-in has not been loaded, you will need to load it before continuing. Follow the instructions in Section GS 8.2.

1. Open a new Excel worksheet and enter the digits 1 through 6 in column A. These digits represent the possible outcomes of rolling a die.

	A
1	1
2	2
3	3
4	4
5	5
6	6

2. At the top of the screen, click **Data** and select **Data Analysis**. Select **Sampling** and click **OK**.

If Data Analysis does not appear as a choice in the Data ribbon, you will need to load the Microsoft Excel Analysis ToolPak add-in. Follow the procedure in Section GS 8.1 before continuing.

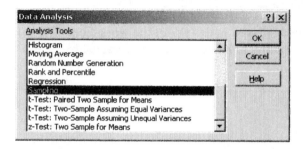

3. Complete the Sampling dialog box as shown below. Click **OK**.

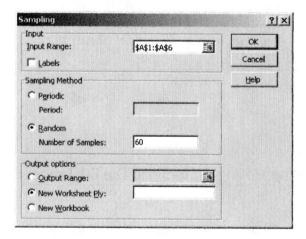

The output will be displayed in a new worksheet. The first 7 entries are shown at the top of the next page. Because the numbers were generated randomly, it is not likely that your output will be exactly the same.

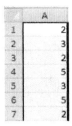

4. To make a tally of the results, first click **Add-Ins** and select **PHStat**. Select **Descriptive Statistics→One-Way Tables & Charts**. Then complete the One-Way Tables & Charts dialog box as shown below. The Type of Data is "Raw Categorical Data." The data are located in cells A1 through A60. There is no label in the first cell (A1). A bar chart will be part of the output. Click **OK**.

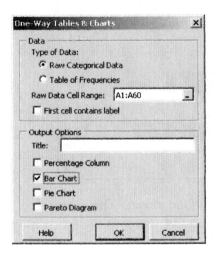

The bar chart is displayed in a sheet named "Bar Chart."

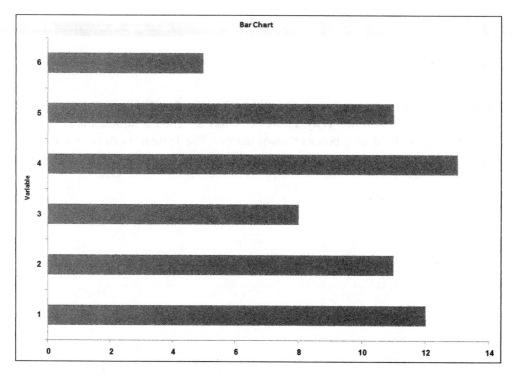

5. Click on the **OneWay Table** sheet tab at the bottom of the screen. This sheet contains a frequency distribution of the outcomes.

	A	B	C
1	One-Way Summary Table		
2			
3	Count of Variable		
4	Variable ▼	Total	
5	1	12	
6	2	11	
7	3	8	
8	4	13	
9	5	11	
10	6	5	
11	Grand Total	60	

6. Click on the **DataCopy** sheet tab at the bottom of the screen. This sheet contains a copy of the generated data set with the word "Variable" inserted in cell A1.

◄

Descriptive Statistics

CHAPTER 2

Section 2.1 Frequency Distributions and Their Graphs

► Example 7 (pg. 48)	Using Technology to Construct Histograms

You will be constructing a histogram for the frequency distribution of the Internet data in Example 2 on page 43.

1. Open worksheet "Internet" in the Chapter 2 folder. These data represent the number of minutes 50 Internet subscribers spent during their most recent Internet session. You will use Excel's Data Analysis to construct a histogram for these data.

If Data Analysis does not appear as a choice in the Data ribbon, you will need to load the Microsoft Excel Analysis ToolPak add-in. Follow the procedure in Section GS 8.1 before continuing.

---

2. Excel's histogram procedure uses grouped data to generate a frequency distribution and a frequency histogram. The procedure requires that you indicate a "bin" for each class. The number that you specify for each bin is actually the upper limit of the class. The upper limits for the Internet data are given on page 41 of your text and are based on a class width of 12. You see that 18 is the upper limit for the first class, 30 is the upper limit for the second class, and so on. The bin for the first class will contain a count of all observations less than or equal to 18, the bin for the second class will contain a count of all observations between 19 and 30, and so on. Enter **Bin** in cell B1 and key in the upper limits in column B as shown at the top of the next page.

	A	B
1	Minutes spent on the internet	Bin
2	50	18
3	40	30
4	41	42
5	17	54
6	11	66
7	7	78
8	22	90

3. Click **Data** and select **Data Analysis**. Select **Histogram**, and click **OK**.

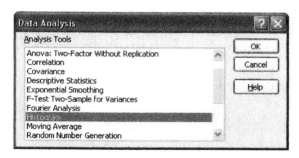

4. Complete the Histogram dialog box as shown below and click **OK**.

To enter the input and bin ranges quickly, follow these steps. First click in the Input Range field of the dialog box. Then, in the worksheet, click and drag from cell A1 to cell A51. You will then see $A$1:$A$51 displayed in the Input Range field. Next click in the Bin Range field. Then click and drag from cell B1 through cell B8 in the worksheet. You will see $B$1:$B$8 displayed in the Bin Range field.

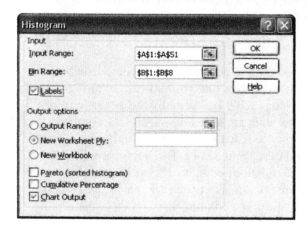

Note the checkmark in the box to the left of Labels. In your worksheet, "Minutes spent on the internet" appears in cell A1, and "Bin" appears in cell B1. Because you included these cells in the Input Range and Bin Range, respectively, you need to let Excel know that these cells contain labels rather than data. Otherwise Excel will attempt to use the information in these cells when constructing the frequency distribution and histogram.

You should see output similar to the output displayed below.

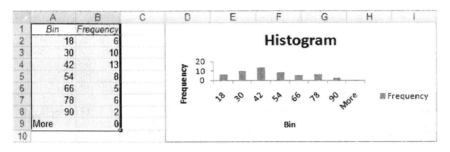

If you want to change the height and width of the chart, begin by clicking anywhere within the figure. Handles appear along the perimeter. You can change the shape of the figure by clicking on a handle and dragging it. You can make the figure small and short or you can make it tall and wide. You can also click within the figure and drag it to move it to a different location on the worksheet.

5. You will now follow steps to modify the histogram so that it is displayed in a more informative and attractive manner. First, make the chart taller so that it is easier to read. To do this, first click within the figure near a border. Dotted handles appear. Click on the center handle on the bottom border of the figure and drag it down a few rows.

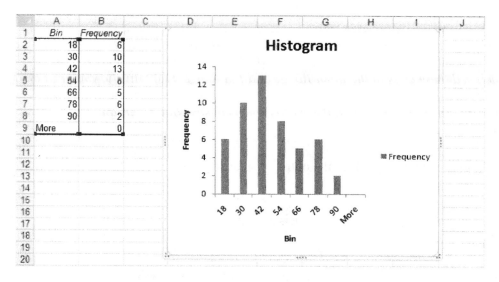

6.   Next remove the space between the vertical bars. **Right-click** on one of the vertical
bars. Select **Format Data Series** from the shortcut menu that appears.

7.   Move the Gap Width arrow all the way to the right for a gap width of 0%. Click
Close.

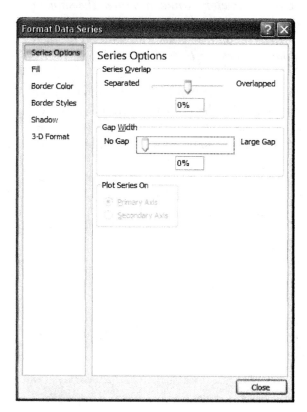

8. Next, change the X-axis values from upper limits to midpoints. The midpoints are displayed in a table on page 43 of your textbook. Enter these midpoints in column C of the Excel worksheet as shown at the top of the next page.

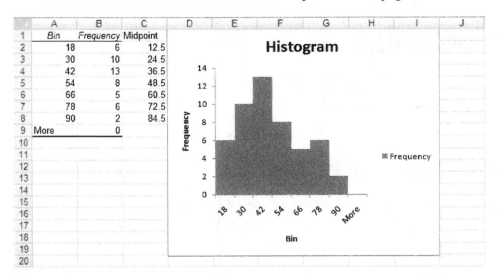

9. **Right-click** on a vertical bar. Click **Select Data** in the shortcut menu that appears.

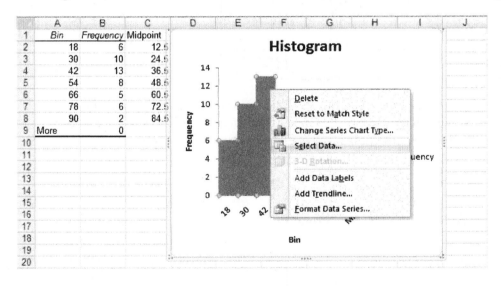

10. Click **Edit** under Horizontal (Category) Axis Labels. Click and drag over the range **C2:C8** to enter the worksheet location of the midpoints into the Axis Labels dialog box. Click **OK**.

11. Click **OK** at the bottom of the Select Data Source dialog box.

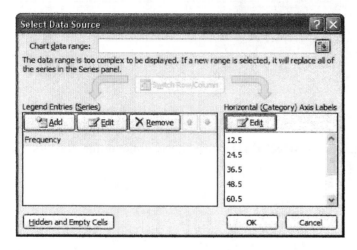

12. Next, you will change the chart title and the x-axis title. Click on the word "Histogram" so that it appears in a box, and change the title to **Internet Usage**. Click on "Bin" and change the x-axis title to **Time Online (in minutes)**.

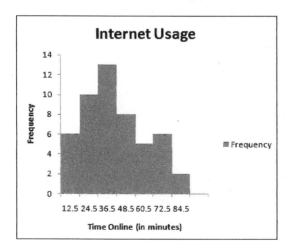

13. Now you will add gridlines. **Right-click** on any number in the y-axis scale. I clicked on 8. Select **Add Major Gridlines** from the menu that appears.

14. Click directly on the **Frequency** legend on the right side of the chart so that it is displayed within a border. Press [**Delete**].

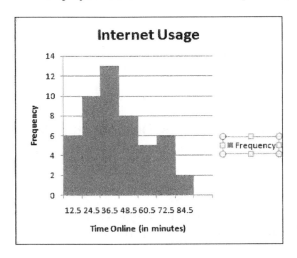

The completed chart is shown below.

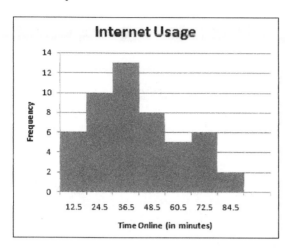

▶ Exercise 29 (pg. 52) | Constructing a Frequency Distribution and Frequency Histogram for the July Sales Data

1.   Open worksheet "Ex2_1-29" in the Chapter 2 folder.

2.   Sort the data so that it is easy to identify the minimum and maximum data entries. In the sorted data set, you see that the minimum sales value is 1000 and that the maximum is 7119.

For instructions on how to sort, see Sections GS 5.1 and GS 5.2.

---

3.   Calculate class width using the formula given in your textbook:

$$\text{Class width} = \frac{\text{Maximum data entry} - \text{Minimum data entry}}{\text{Number of classes}}$$

The exercise instructs you to use 6 for the number of classes.

$$\text{Class width} = \frac{7119 - 1000}{6} = 1019.83 \text{ . Round up to } 1020.$$

4. The textbook instructs you to use the minimum data entry as the lower limit of the first class. To find the remaining lower limits, add the class width of 1020 to the lower limit of each previous class. To do this, begin by entering **Lower Limit** in cell B1 of the worksheet and **1000** in cell B2. You will now use a formula to compute the remaining lower limits, and you will have Excel do these calculations for you. In cell B3, enter the formula **=B2+1020** as shown below. Press **[Enter]**.

	A	B	C
1	July sales	Lower Limit	
2	1000	1000	
3		1030	=B2+1020

5. Click in cell **B3** where 2020 now appears and copy the formula in cell B3 to cells B4 through B8. Because 7119 is the maximum data entry, you don't need 7120 for the histogram. However, you will be using 7120 when calculating the upper limit for the last class interval.

	A	B	C
1	July sales	Lower Limit	
2	1000	1000	
3	1030	2020	
4	1077	3040	
5	1355	4060	
6	1500	5080	
7	1512	6100	
8	1643	7120	

6. Enter **Upper Limit** in cell C1 of the worksheet. The textbook says that the upper limit is equal to one less than the lower limit of the next higher class. You will use a formula to compute the upper limits, and you will have Excel do the computations for you. Click in cell **C2** and enter the formula **=B3-1** as shown below. Press **[Enter]**.

	A	B	C	D
1	July sales	Lower Limit	Upper Limit	
2	1000	1000	=B3-1	
3	1030	2020		

7. Copy the formula in cell C2 (where 2019 now appears) to cells C3 through C7. These are the upper limits that you will use as bins for constructing the histogram chart.

	A	B	C	D
1	July sales	Lower Limit	Upper Limit	
2	1000	1000	2019	
3	1030	2020	3039	
4	1077	3040	4059	
5	1355	4060	5079	
6	1500	5080	6099	
7	1512	6100	7119	

8. Click **Data** and select **Data Analysis**. Select **Histogram** and click **OK**.

If Data Analysis does not appear as a choice in the Data ribbon, you will need to load the Microsoft Excel Analysis ToolPak add-in. Follow the procedure in Section GS 8.1 before continuing

---

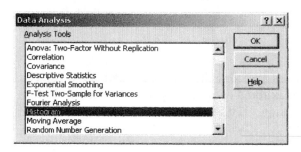

9. Complete the fields in the Histogram dialog box as shown below. The July sales data are located in cells A1 through A23 of the worksheet. The bins (upper limits) are located in cells C1 through C7. The top cells in these ranges are labels—"July sales" and "Upper Limit," so click in the Labels box to place a checkmark there. The output will be placed in a new Excel workbook. The output will include a chart. Click **OK**.

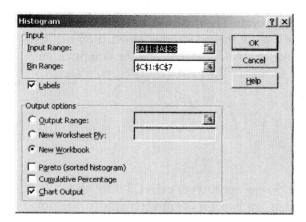

You should see output similar to the output displayed below.

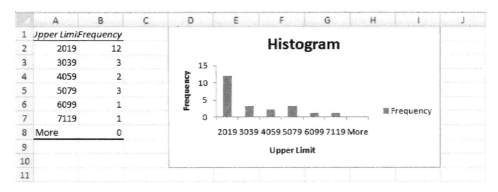

10. You will now follow steps to modify the histogram so that it is displayed in a more
     accurate and informative manner. First, make the chart taller so that it is easier to
     read. To do this, first click within the figure near a border. Dotted handles appear.
     Click on the center handle on the bottom border of the figure and drag it down a few
     rows. Next, click on the handle on the right side and drag it a couple of columns to
     the right.

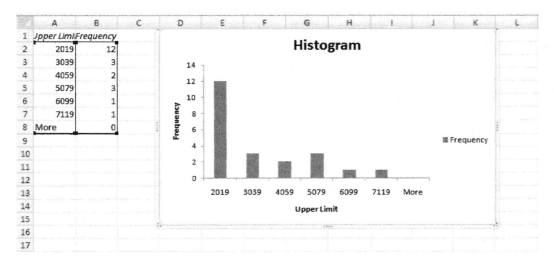

11. Remove the space between the vertical bars. **Right-click** on one of the vertical bars. Select **Format Data Series** from the shortcut menu that appears.

12. Move the Gap Width arrow all the way to **No Gap**. Click **Close**.

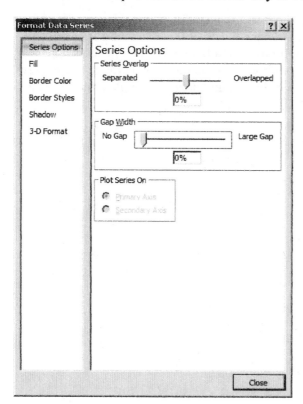

13. Delete the word "More" from the X axis. To do this, **right-click** on a vertical bar and select **Select Data** from the shortcut menu that appears.

14. Click **Edit** under Legend Entries (Series).

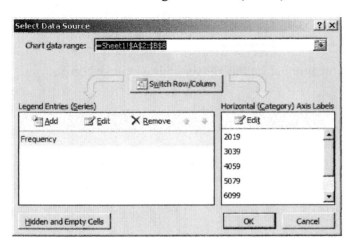

15. You do not want to include row 8 of the frequency distribution because that is the row containing information related to the "More" category. Change the 8 to 7 in the **Series values** field so that it reads **=Sheet1!$B$2:$B$7**. Click **OK**. Click **OK** in the Select Data Source dialog box.

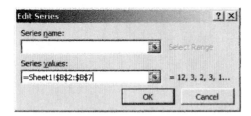

16. You will now revise the chart and axis titles. Click on the "Histogram" chart title and change it to **July Sales**. Click on "Upper Limit" and change it to **Dollars**.

17. To remove the legend, click on the **Frequency** legend at the right and press [**Delete**].

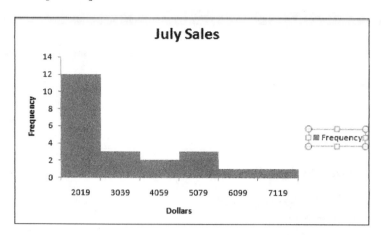

18. To add gridlines, **right-click** on a number on the Y-axis scale. I clicked on 8. Select **Add Major Gridlines**.

	Delete
	Reset to Match Style
A	Font...
	Change Chart Type...
	Select Data...
	3-D Rotation...
	Add Major Gridlines
	Add Minor Gridlines
	Format Axis...

The completed histogram is shown below.

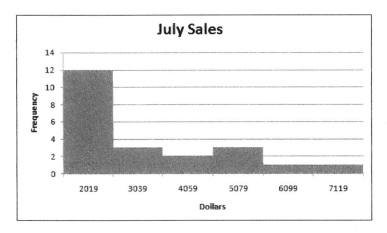

Section 2.2 More Graphs and Displays

▶ Example 1 (pg. 55)	Constructing a Stem-and-Leaf Plot of Numbers of Text Messages Sent

If the PHStat2 add-in has not been loaded, you will need to load it before continuing. Follow the instructions in Section GS 8.2.

1. Open worksheet "Text Messages" in the Chapter 2 folder.

2. At the top of the screen, click **Add-Ins** and select **PHStat**. Select→**Descriptive Statistics→Stem-and-Leaf Display**.

3. Complete the entries in the Stem-and-Leaf Display dialog box as shown at the top of the next page. Click **OK**.

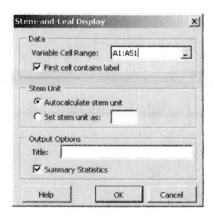

The output is displayed in a new sheet named "StemLeafPlot."

	A	B	C	D	E
1				Stem-and-Leaf Display	
2					
3				Stem unit 10	
4					
5		Statistics		7	8
6	Sample Size	50		8	
7	Mean	125.14		9	
8	Median	123.5		10	5 8 9 9 9
9	Std. Deviation	14.53919		11	2 2 2 3 4 6 7 8 8 8 9 9 9
10	Minimum	78		12	1 1 2 2 2 3 4 4 6 6 6 6 6 9 9
11	Maximum	159		13	0 2 3 3 4 9 9
12				14	0 2 4 5 5 7 8
13				15	5 9
14					

◀

▶ Example 3 (pg. 57)	Constructing a Dot Plot

If the PHStat2 add-in has not been loaded, you will need to load it before continuing. Follow the instructions in Section GS 8.2.

1. Open worksheet "Text Messages" in the Chapter 2 folder.

2. At the top of the screen, click **Add-Ins** and select **PHStat**. Select **Descriptive Statistics→Dot Scale Diagram.**

3. Complete the entries in the Dot Scale Diagram dialog box as shown below. Click
    OK.

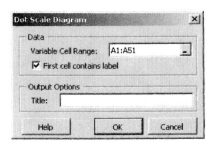

The output is displayed in a new sheet named "Dot Scale."

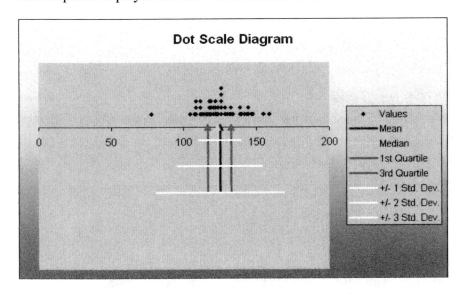

◀

▶ Example 4 (pg. 58)	Constructing a Pie Chart

1. Open a new Excel worksheet and enter the motor vehicle frequency table as shown
    at the top of the next page.

	A	B
1	Vehicle type	Killed
2	Cars	18440
3	Trucks	13778
4	Motorcycles	4552
5	Other	823

2. Highlight cells **A2:B5**. Then click **Insert** at the top of the screen and select **Pie**. Select the leftmost diagram in the 2-D Pie row.

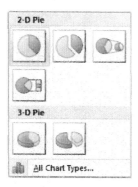

3. Click **Layout** at the top of the screen and select **Chart Title**. Select **Above Chart**.

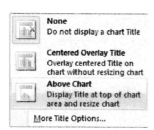

4. Replace "Chart Title" with **Motor Vehicle Occupants Killed in 2005**.

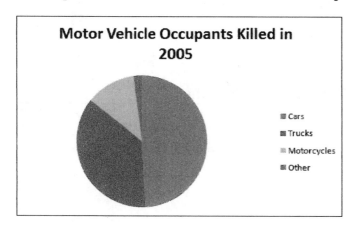

5. **Right-click** directly on the pie chart and select **Add Data Labels**.

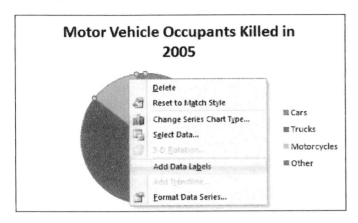

6. **Right-click** on a data label. I clicked on 823. Select **Format Data Labels**.

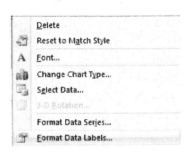

7. Select **Percentage**. Click **Close**.

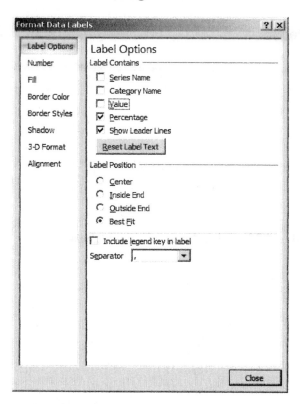

8. You will now move the pie chart to a new worksheet. **Right-click** in the white space around the chart. Select **Move Chart**.

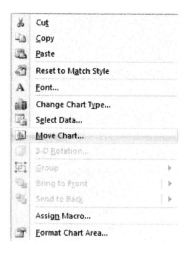

9. Select **New sheet**. Click **OK**.

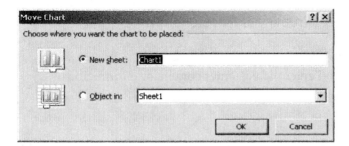

The completed pie chart is shown below.

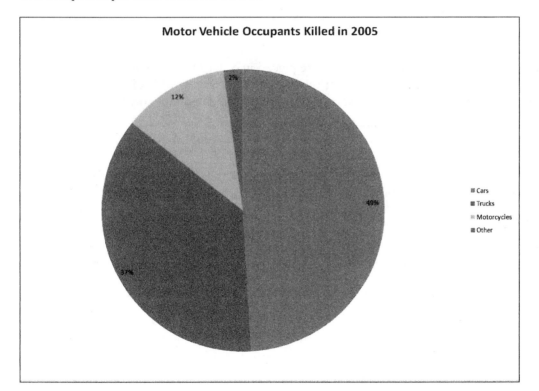

| ► Example 5 (pg. 59) | Constructing a Pareto Chart |

---

1. Open a new Excel worksheet, and enter the customer complaint information as shown below.

	A	B
1	Source	Frequency
2	Home furnishing	7792
3	Computer sales & service	5733
4	Auto dealers	14668
5	Auto repair	9728
6	Dry cleaning	4649

2. In a Pareto chart, the vertical bars are placed in order of decreasing height. Excel's Chart function will display the information in the order in which it appears in the worksheet. So, you need to sort the information in descending order by Frequency.

For instructions on how to sort, see Sections GS 5.1 and GS 5.2.

	A	B
1	Source	Frequency
2	Auto dealers	14668
3	Auto repair	9728
4	Home furnishing	7792
5	Computer sales & service	5733
6	Dry cleaning	4649

3. After sorting the data, highlight cells **A1:B6** and click **Insert** at the top of the screen.

4. Select a **Column** chart and select the leftmost figure in the 2-D column row.

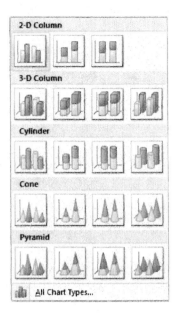

5. Next you will add an X-axis title. Click **Layout** at the top of the screen and select **Axis Titles→Primary Horizontal Axis→Title Below Axis**. Replace "Axis Title" with **Type of Business**.

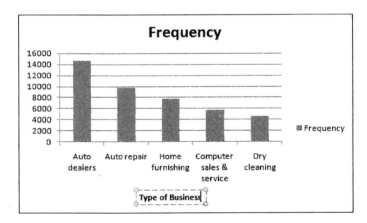

6. Now you will add the Y-axis title. Click **Layout** at the top of the screen and select **Axis Titles→Primary Vertical Axis Title →Vertical Title**. Replace "Axis Title" with **Frequency**.

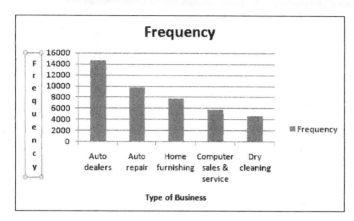

7. To remove the legend, click on the **Frequency** legend at the right and press **[Delete]**.

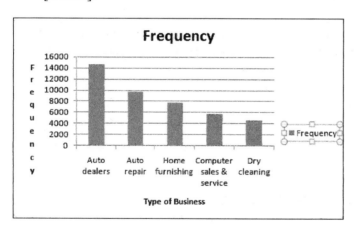

Your chart should look similar to the one shown below.

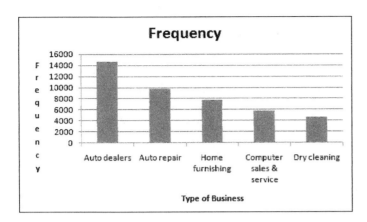

▶ Example 7 (pg. 61)   Constructing a Time Series Chart

1. Open worksheet "Cellular Phones" in the Chapter 2 folder.

2. Highlight cells **A3:B13**. At the top of the screen, click **Insert** and select a **Scatter** chart.

	A	B	C
1	Year	Subscribers	Avg. Bill $
2			
3	1995	33.8	51.00
4	1996	44.0	47.70
5	1997	55.3	42.78
6	1998	69.2	39.43
7	1999	86.0	41.24
8	2000	109.5	45.27
9	2001	128.4	47.37
10	2002	140.8	48.40
11	2003	158.7	49.91
12	2004	182.1	50.64
13	2005	207.9	49.98

3.   Select the leftmost scatter diagram in the top row.

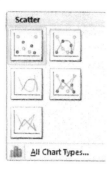

4.   At the top of the screen, click **Design** and select **Layout 1** so that you can add a chart title and axis titles.

5.   Replace "Chart Title" with **Cellular Telephone Subscribers by Year**. Replace the Y-Axis Title with **Subscribers (in millions)**. Replace the X-Axis Title with **Year**.

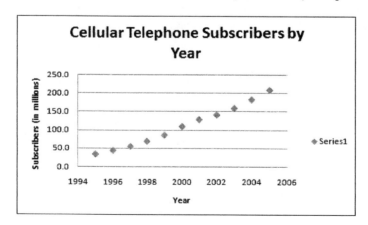

6. To remove the legend, click on the **Series1** legend to the right of the chart. Press [**Delete**].

Your scatter plot should now look similar to the one shown below.

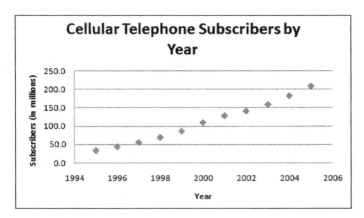

Section 2.3 Measures of Central Tendency

| ▶ Example 6 (pg. 70) | Comparing the Mean, Median, and Mode |

If the PHStat2 add-in has not been loaded, you will need to load it before continuing. Follow the instructions in Section GS 8.2.

1. Open worksheet "Ages" in the Chapter 2 folder. The instructions for Try It Yourself 6 at the bottom of page 70 tell you to remove the data entry of 65 years. Because 65 is the bottom entry in the data set, you do not need to delete it. Instead, you will not include it in the input range.

2. At the top of the screen, click **Data** and select **Data Analysis**. Select **Descriptive Statistics** and click **OK**.

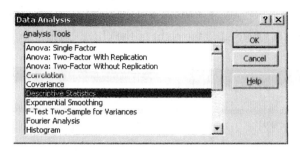

3. Complete the Descriptive Statistics dialog box as shown at the top of the next page. The entry in the **Input Range** field is the worksheet location of the age data. The checkmark to the left of **Labels in First Row** lets Excel know that the entry in cell A1 is a label and is not to be included in the computations. The output will be placed in a **New Worksheet**. The checkmark to the left of **Summary statistics** requests that the output include summary statistics for the specified set of data. Click **OK**.

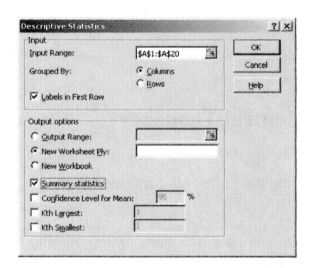

4. You will want to increase the width of column A of the output so that you can read the labels. Your output should be similar to the output shown at the top of the next page. The mean of the sample is 21.58, the median is 21, and the mode is 20.

Be careful when using the value of the mode reported in the Descriptive Statistics output. If there is a tie for the mode, Excel reports only the first modal value that occurs in the data set. Therefore, it is always a good idea to construct a frequency distribution table to go along with the Descriptive Statistics output.

	A	B
1	Ages of students in class	
2		
3	Mean	21.57895
4	Standard Error	0.327276
5	Median	21
6	Mode	20
7	Standard Deviation	1.426565
8	Sample Variance	2.035088
9	Kurtosis	-1.27106
10	Skewness	0.335563
11	Range	4
12	Minimum	20
13	Maximum	24
14	Sum	410
15	Count	19

5. Construct a frequency distribution table to see if the data set is unimodal or multimodal. Go back to the worksheet containing the age data. To do this, click on the **Sheet1** tab near the bottom of the screen. Click **Add-Ins** at the top of the screen and select **PHStat**. Select **Descriptive Statistics→One-Way Tables & Charts**.

6. Complete the One-Way Tables & Charts dialog box as shown below. Click **OK**.

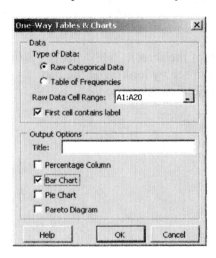

The bar chart is displayed in a worksheet named "Bar Chart."

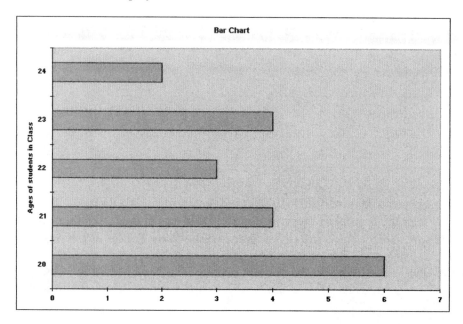

7. Click on the **OneWay Table** sheet tab at the bottom of the screen. This sheet contains a frequency distribution of the ages. You can see by looking at the bar chart and the frequency distribution that the age distribution is unimodal. The modal value of 20 has a frequency of six.

	A	B
1	One-Way Summary Table	
2		
3	Count of Ages of students in class	
4	Ages of students in class ▾	Total
5	20	6
6	21	4
7	22	3
8	23	4
9	24	2
10	Grand Total	19

Section 2.4 Measures of Variation

| ▶ Example 5 (pg. 86) | Using Technology to Find the Standard Deviation |

1. Open worksheet "Rent Rates" in the Chapter 2 folder.

2. Click **Data** and select **Data Analysis**. Select **Descriptive Statistics** and click **OK**.

If Data Analysis does not appear as a choice in the Data ribbon, you will need to load the Microsoft Excel Analysis ToolPak add-in. Follow the procedure in Section GS 8.1 before continuing.

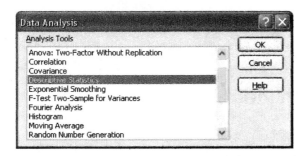

3. Complete the Descriptive Statistics dialog box as shown below. Click **OK**.

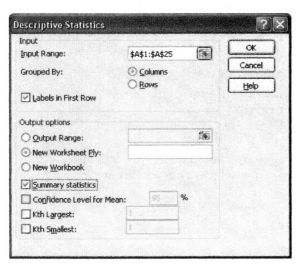

You will want to make Column A of the output wider so that you can read the labels. Your Descriptive Statistics output should be similar to the output shown below. The mean is 33.73 and the standard deviation is 5.09.

	A	B
1	Office rental rates	
2		
3	Mean	33.72917
4	Standard Error	1.038864
5	Median	35.375
6	Mode	37
7	Standard Deviation	5.089373
8	Sample Variance	25.90172
9	Kurtosis	-0.74282
10	Skewness	-0.70345
11	Range	16.75
12	Minimum	23.75
13	Maximum	40.5
14	Sum	809.5
15	Count	24

◀

Section 2.5 Measures of Position

▶ Example 2 (pg. 103)	Using Technology to Find Quartiles

1. Open worksheet "Tuition" in the Chapter 2 folder. Key in labels for the quartiles as shown below. Then click in cell **D2** to place the first quartile there.

	A	B	C	D
1	Tuition costs (thousands of dollars)		Quartile	
2	23		1	
3	25		2	
4	30		3	

2. You will be using the QUARTILE function to obtain the first, second, and third quartiles for the tuition data. At the top of the screen, click **Formulas** and select **Insert Function**.

3.   Select the **Statistical** category. Select the **QUARTILE** function. Click **OK**.

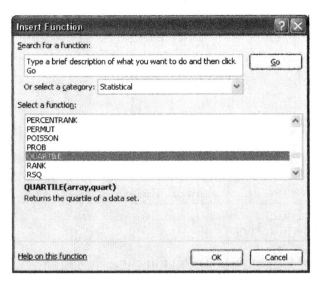

4.   Complete the QUARTILE dialog box as shown below. Click **OK**.

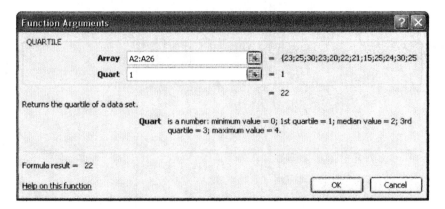

5.   Click in cell **D3** to place the second quartile there.

6.   At the top of the screen, click **Formulas** and select **Insert Function**.

7.   Select the **Statistical** category and the **QUARTILE** function. Click **OK**.

8.   Complete the QUARTILE dialog box as shown below. Click **OK**.

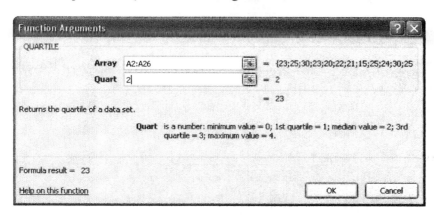

9.   Click in cell **D4** to place the third quartile there.

10.   At the top of the screen, click **Formulas** and select **Insert Function**.

11.   Select the **Statistical** category and the **QUARTILE** function. Click **OK**.

12.   Complete the QUARTILE dialog box as shown below. Click **OK**.

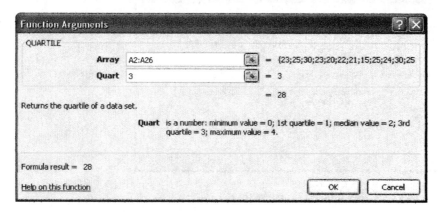

Your output should look similar to the output displayed below.

	A	B	C	D
1	Tuition costs (thousands of dollars)		Quartile	
2	23		1	22
3	25		2	23
4	30		3	28

> ► Example 4 (pg. 105)   Drawing a Box-and-Whisker Plot
> _____

_____

*If the PHStat2 add-in has not been loaded, you will need to load it before continuing.
Follow the instructions in Section GS 8.2.*

_____

1. Open a new Excel worksheet and enter the test scores of 15 employees enrolled in a
    CPR training course as shown below.

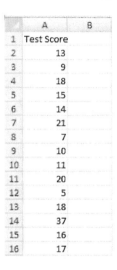

	A	B
1	Test Score	
2	13	
3	9	
4	18	
5	15	
6	14	
7	21	
8	7	
9	10	
10	11	
11	20	
12	5	
13	18	
14	37	
15	16	
16	17	

2. At the top of the screen, click **Add-Ins** and select **PHStat**. Select **Descriptive
    Statistics→Box-and-Whisker Plot**.

3. Complete the Box-and-Whisker Plot dialog box as shown below. Click **OK**.

The Five-Number Summary is displayed in a worksheet named "FiveNumbers."

	A	B
1	CPR Training Test Score	
2		
3	Five-number Summary	
4	Minimum	5
5	First Quartile	10
6	Median	15
7	Third Quartile	18
8	Maximum	37

The box-and-whisker plot is displayed in a worksheet named "BoxWhiskerPlot."

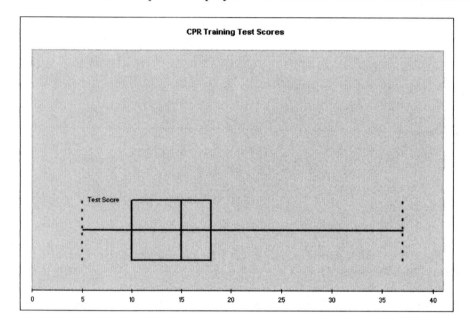

Technology

▶ Exercises 1 & 2   (pg. 123)   Finding the Sample Mean and the Sample Standard Deviation

1. Open worksheet "Tech2" in the Chapter 2 folder.

2. Click **Data** and select **Data Analysis**. Select **Descriptive Statistics** and click OK.

If Data Analysis does not appear as a choice in the Data ribbon, you will need to load the Microsoft Excel Analysis ToolPak add-in. Follow the procedure in Section GS 8.1 before continuing.

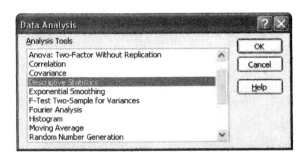

3. Complete the Descriptive Statistics dialog box as shown below. Click **OK**.

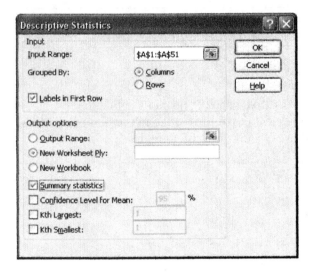

You will want to adjust the width of column A of the output so that you can read the labels. Your Descriptive Statistics output should be similar to the output shown at the top of the next page. The sample mean is 2270.54 and the sample standard deviation is 653.1822.

	A	B
1	Monthly production (in pounds)	
2		
3	Mean	2270.54
4	Standard Error	92.37391
5	Median	2207
6	Mode	2207
7	Standard Deviation	653.1822
8	Sample Variance	426647
9	Kurtosis	0.567664
10	Skewness	0.549267
11	Range	3138
12	Minimum	1147
13	Maximum	4285
14	Sum	113527
15	Count	50

◀

▶ Exercises 3 & 4 (pg. 123)	Constructing a Frequency Distribution and a Frequency Histogram

1. Open worksheet "Tech2" in the Chapter 2 folder.

2. Sort the data in ascending order. In the sorted data set you can see that the minimum production is 1147 pounds. Scroll down to find the maximum production. The maximum production, displayed in cell A51, is 4285 pounds.

For instructions on how to sort, refer to Sections GS 5.1 and GS 5.2.

	A
1	Monthly production (in pounds)
2	1147
3	1230
50	3512
51	4285

3. Enter **Lower Limit** in cell D1 of the worksheet. The textbook tells you to use the minimum value as the lower limit of the first class. Enter **1147** in cell D2. You will calculate the remaining lower limits by adding the class width of 500 to the lower limit of each previous class. You will use a formula to do these computations in the Excel worksheet. Click in cell **D3** and key in **=D2+500** as shown below. Press **[Enter]**.

	A	B	C	D
1	Monthly production (in pounds)			Lower Limit
2	1147			1147
3	1230			=D2+500

4. Click in cell **D3** (where 1647 now appears) and copy the contents of cell D3 to cells D4 through D9. Because the maximum milk production is 4285 pounds, you have calculated one more lower limit than is needed for the histogram. The value of 4647, however, will be used when calculating the upper limit of the last class.

	A	B	C	D
1	Monthly production (in pounds)			Lower Limit
2	1147			1147
3	1230			1647
4	1258			2147
5	1294			2647
6	1319			3147
7	1449			3647
8	1619			4147
9	1647			4647

5. Enter **Upper Limit** in cell E1. The upper limit is equal to one less than the lower limit of the next higher class. To do these calculations, you will enter a formula in the Excel worksheet. Click in cell E2 and enter the formula **=D3-1** as shown in the worksheet below. Press [**Enter**].

	A	B	C	D	E	F
1	Monthly production (in pounds)			Lower Limit	Upper Limit	
2	1147			1147	=D3-1	
3	1230			1647		

6. Copy the formula in E2 (where 1646 now appears) to cells E3 through E8. You will use these upper limits for the bins when you construct the histogram chart.

	A	B	C	D	E	F
1	Monthly production (in pounds)			Lower Limit	Upper Limit	
2	1147			1147	1646	
3	1230			1647	2146	
4	1258			2147	2646	
5	1294			2647	3146	
6	1319			3147	3646	
7	1449			3647	4146	
8	1619			4147	4646	

7. At the top of the screen, click **Data** and select **Data Analysis**. Select **Histogram** and click **OK**.

If Data Analysis does not appear as a choice in the Data ribbon, you will need to load the Microsoft Excel Analysis ToolPak add-in. Follow the procedure in Section GS 8.1 before continuing.

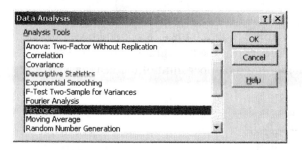

8. Complete the fields in the histogram dialog box as shown below. Click **OK**.

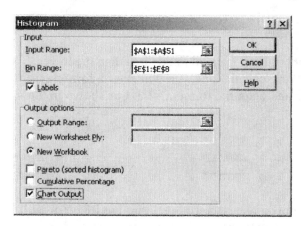

9. You will now follow steps to modify the histogram so that it is presented in a more informative manner. Begin by making the chart taller so that it is easier to read. To do this, first click within the figure near a border. Dotted handles appear. Click on the center handle at the bottom border of the figure and drag it down a few rows.

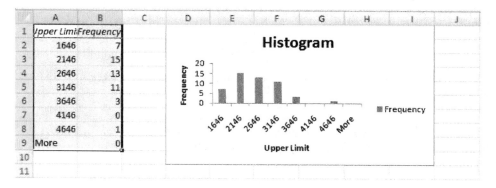

10. To remove the "More" category from the X axis first **right-click** on a vertical bar in the histogram and select **Source Data** from the shortcut menu that appears.

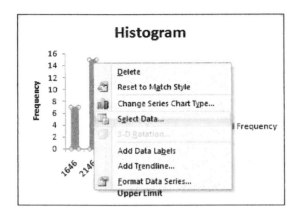

11. Click **Edit** under Horizontal (Category) Axis Labels.

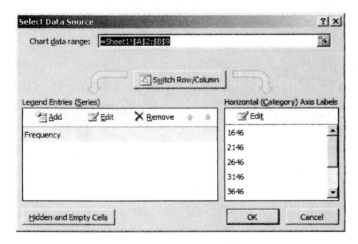

12. "More" appears in cell A9 of the worksheet. To exclude the information in row 9 from the chart, change the 9 to 8 in the Axis label range window. Click **OK**. Click **OK** in the Select Data Source dialog box.

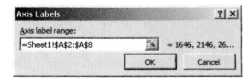

13. Next, change the chart title and the X-axis title. Click on "Histogram" and enter **Monthly Milk Production of 50 Holstein Dairy Cows**. Click on "Upper Limit" and enter **Pounds**,

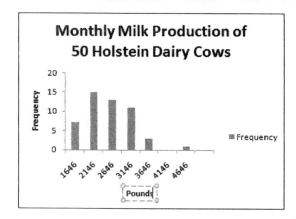

14. To remove the legend, click on the **Frequency** legend at the right and press [**Delete**].

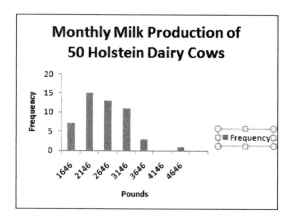

15. To add gridlines, **right-click** directly on a number in the Y-axis. Select I clicked on 20. Select **Add Major Gridlines**.

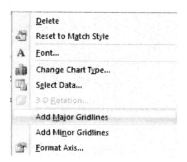

16. To remove the space between the vertical bars, **right-click** on one of the vertical bars and select **Format Data Series** from the shortcut menu that appears.

17. Move the Gap Width arrow all the way to the left so that 0% shows in the window. Click **Close**.

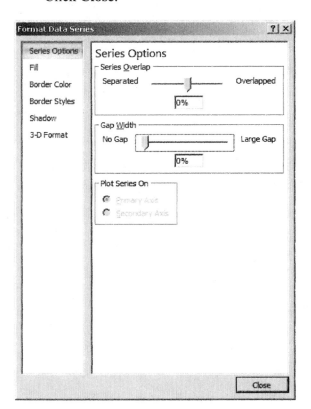

Your completed histogram should look similar to the one displayed at the top of the next page.

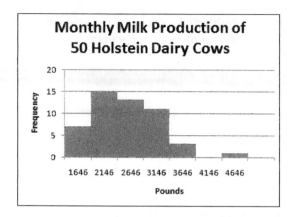

Probability

Section 3.2 Conditional Probability and the Multiplication Rule

> ► Exercise 36 (pg. 159)   Simulating the "Birthday Problem"

Part d of Exercise 36 asks you to use a technology tool to simulate the "Birthday Problem." You will be generating 24 random numbers between 1 and 365.

If the PHStat2add-in has not been loaded, you will need to load it before continuing. Follow the instructions in Section GS 8.2.

---

1. Open a new Excel worksheet.

2. You will be entering the numbers 1 through 365 in the worksheet. To do this, you will fill column A with a number series. Begin by entering the numbers 1 and 2 in column A as shown below. You are starting a number series that will increase in increments of one.

A
1
2

3. You will now complete the series all the way to 365. Begin by clicking in cell A1 and dragging down to cell A2 so that both cells are highlighted as shown below.

4. Move the mouse pointer in cell A2 to the bottom right corner of that cell so that the white plus sign turns into a black plus sign. This black plus sign is called the "fill handle." While the fill handle is displayed, hold down on the left mouse button and

drag down column A until you reach cell A365. Release the mouse button. You should see the numbers from 1 to 365 in column A.

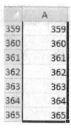

5. At the top of the screen, click **Data** and select **Data Analysis**. Select **Sampling** and click **OK**.

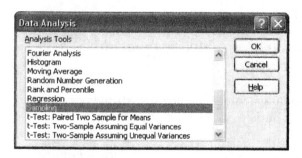

6. Complete the Sampling dialog box as shown below. Click **OK**.

You will be generating 10 different samples of 24 numbers. The first set of 24 numbers will be placed in column B, the second in column C, etc.

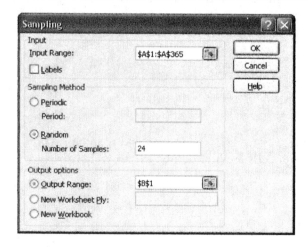

7. The 24 numbers, generated randomly with replacement, are displayed in column B. (Because the numbers were generated randomly, it is not likely that your output will be exactly the same.) Construct a frequency distribution to see if there are any repetitions. Select **Add-Ins** and **PHStat**. Select **Descriptive Statistics→One-Way Tables & Charts**.

	A	B
1	1	129
2	2	270
3	3	257
4	4	147
5	5	17
6	6	332
7	7	282
8	8	350
9	9	333
10	10	226
11	11	182
12	12	350
13	13	66
14	14	191
15	15	346
16	16	45
17	17	311
18	18	100
19	19	318
20	20	200
21	21	15
22	22	333
23	23	75
24	24	12

8. Complete the One-Way Tables & Charts dialog box as shown below. Be sure to remove the checkmark in the box to the left of **First cell contains label**. Click **OK**.

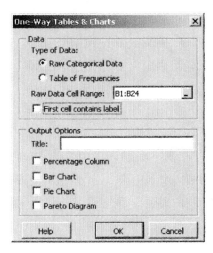

9. The frequency distribution table is displayed in a worksheet named "OneWayTable." In this example, you can see that 333 has a frequency of 2.

	A	B	C
1	One-Way Summary Table		
2			
3	Count of Variable		
4	Variable ▾	Total	
5	12	1	
6	15	1	
7	17	1	
8	45	1	
9	66	1	
10	75	1	
11	100	1	
12	129	1	
13	147	1	
14	182	1	
15	191	1	
16	200	1	
17	226	1	
18	257	1	
19	270	1	
20	282	1	
21	311	1	
22	318	1	
23	332	1	
24	333	2	
25	346	1	
26	350	2	
27	Grand Total	24	

10. Generate the second sample. Go back to the sheet containing the numbers 1 through 365 by clicking on the **Sheet1** tab near the bottom of the screen. Then click **Data** and select **Data Analysis**. Select **Sampling** and click **OK**. Complete the Sampling dialog box as shown below. The second set of randomly generated numbers will be placed in column C. Click **OK**.

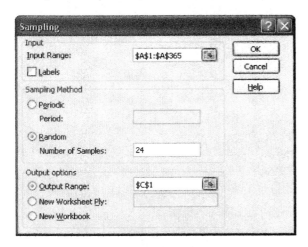

11. Construct a frequency distribution to see if there are any repetitions. Select **Add-Ins** and **PHStat**. Select **Descriptive Statistics→One-Way Tables & Charts**. Complete the One-Way Tables & Charts dialog box as shown below. Click **OK**.

12. The frequency distribution table is displayed in a worksheet named "OneWayTable2." In this example, you can see that there are no repetitions.

	A	B	C
1	One-Way Summary Table		
2			
3	Count of Variable		
4	Variable ▾	Total	
5	12	1	
6	16	1	
7	33	1	
8	34	1	
9	67	1	
10	86	1	
11	95	1	
12	142	1	
13	152	1	
14	175	1	
15	177	1	
16	181	1	
17	186	1	
18	244	1	
19	247	1	
20	249	1	
21	252	1	
22	286	1	
23	297	1	
24	298	1	
25	306	1	
26	316	1	
27	351	1	
28	365	1	
29	Grand Total	24	

13. Repeat the appropriate steps until you have generated 10 different samples of 24 numbers.

Section 3.4 Additional Topics in Probability and Counting

► Example 1 (pg. 172)	Finding the Number of Permutations of n Objects

1. You are asked to determine how many different final standings are possible in the Central Division. Open a new Excel worksheet and click in cell **A1** to place the output there. You will be using the PERMUT function to calculate the number of permutations.

2. Click **Formulas** and select **Insert Function**.

3. Select the **Statistical** category. Select the **PERMUT** function. Click **OK**.

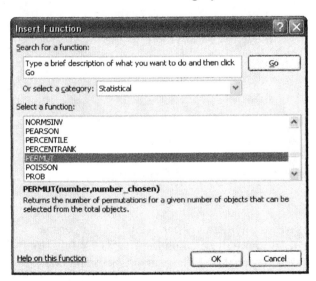

4. Complete the PERMUT dialog box as shown below. There are six teams in the National League Central Division. The number of different final standings is equal to 6! Click **OK**

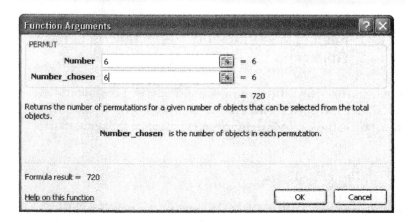

The output is shown below. There are 720 possible different final standings.

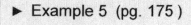

◄

► Example 5 (pg. 175)	Finding the Number of Combinations

1. You are asked to calculate how many ways the manager can form a three-person advisory committee from 20 employees. You will be using the COMBIN function to calculate the number of combinations. Open a new Excel worksheet and click in cell **A1** to place the output there.

2. Click **Formulas** and select **Insert Function**.

3. Select the **Math & Trig** category. Select the **COMBIN** function. Click **OK**.

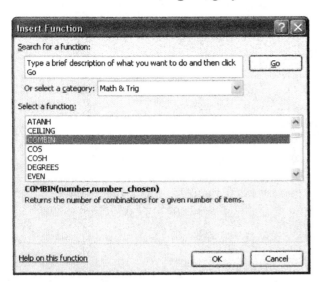

4. Complete the COMBIN dialog box as shown below. Click **OK**.

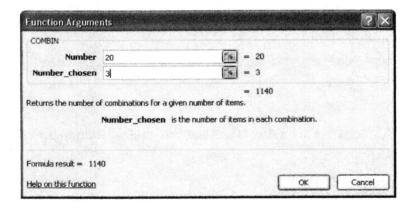

The output is shown below. There are 1,140 different combinations.

Technology

| ▶ Exercise 3 (pg. 191) | Selecting Randomly a Number from 1 to 11 |

If the PHStat2 add-in has not been loaded, you will need to load it before continuing. Follow the instructions in Section GS 8.2.

You will be given the steps to follow to complete Exercise 3 and Exercise 3B.

1. You will first select randomly a number between 1 and 11. Open a new Excel worksheet. Enter the numbers 1 through 11 in cells A1 through A11 as shown below.

2. At the top of the screen, click **Data** and select **Data Analysis**. Select **Sampling**, and click **OK**.

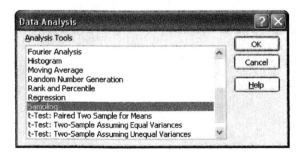

3. Complete the Sampling dialog box as shown below. Click **OK**.

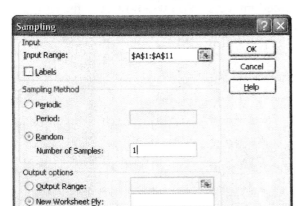

4. The output is displayed in a new sheet. The number 2 was selected. Because the number was selected randomly, your output may not be the same. You will now select randomly 100 integers from 1 to 11. Click the **Sheet1** tab near the bottom of the screen to go back to the worksheet displaying the numbers from 1 to 11.

5. Click **Data** and select **Data Analysis**. Select **Sampling** and click **OK**.

6. Complete the Sampling dialog box as shown below. Click **OK**.

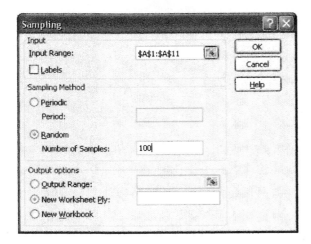

7. The output is displayed in a new sheet. You will need to scroll down to see the entire listing of the 100 randomly generated integers. You will now tally the results. Select **Add-Ins** and **PHStat**. Select **Descriptive Statistics→One-Way Tables & Charts**.

	A
1	11
2	6
3	11
4	1
5	3
6	4
7	1

8. Complete the One-Way Tables & Charts dialog box as shown below. Click **OK**.

Be sure to remove the checkmark next to First cell contains label. Otherwise, your summary table will include only 99 observations.

The output for Exercise 3B is displayed below. Your output should have the same format. Because the numbers were generated randomly, however, it is not likely that your numbers will be exactly the same.

	A	B	C
1	One-Way Summary Table		
2			
3	Count of Variable		
4	Variable ▾	Total	
5	1	12	
6	2	12	
7	3	9	
8	4	10	
9	5	11	
10	6	13	
11	7	6	
12	8	8	
13	9	6	
14	10	7	
15	11	6	
16	Grand Total	100	

◀

▶ **Exercise 5 (pg. 191)** | Composing Mozart Variations with Dice

---

If the PHStat2 add-in has not been loaded, you will need to load it before continuing. Follow the instructions in Section GS 8.2.

You will first be given the steps to follow to complete Exercise 5 and then the steps to follow to complete Exercise 5B. In Exercise 5, you will select randomly two integers from 1, 2, 3, 4, 5, and 6. This simulates tossing two six-sided dice one time. You will then sum the two integers and subtract 1 from the sum. This is Mozart's procedure (described at the top of page 191) for selecting a musical phrase from 11 different choices.

1. To select randomly two integers between 1 and 6, begin by opening a new Excel worksheet and entering the numbers 1 through 6 in column A as shown at the top of the next page.

	A
1	1
2	2
3	3
4	4
5	5
6	6

2. At the top of the screen, click **Data** and select **Data Analysis**. Select **Sampling** and click **OK**.

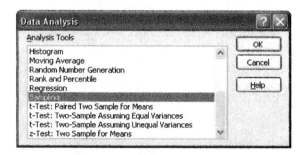

3. Complete the Sampling dialog box as shown below. Click **OK**.

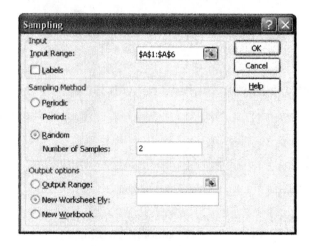

4. The output is displayed in a new worksheet. The numbers 2 and 4 were selected. Mozart's instructions are to sum the numbers and subtract 1. The result is 5. Therefore, musical phrase number 5 would be selected for the first bar of Mozart's minuet.

	A
1	2
2	4

5. Exercise 5B asks you to select 100 integers between 1 and 11. This simulates, 100 times, Mozart's procedure of tossing two die, finding the sum, and subtracting 1. You are also asked to tally the results. Begin by entering the numbers 1 through 11 in a new worksheet as shown below.

	A
1	1
2	2
3	3
4	4
5	5
6	6
7	7
8	8
9	9
10	10
11	11

6. Click **Data** and select **Data Analysis**. Select **Sampling** and click **OK**.

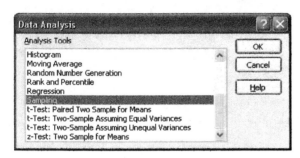

7. Complete the Sampling dialog box as shown below. Click **OK**.

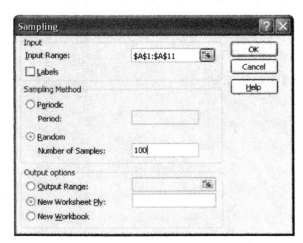

8.   The output is displayed in a new sheet. You need to scroll down to see the entire listing of the 100 randomly generated integers. You will now tally the results. Select **Add-Ins** and **PHStat**. Select **Descriptive Statistics→One-Way Tables & Charts**.

	A
1	5
2	11
3	8
4	11
5	1
6	7
7	5

9.   Complete the One-Way Tables and Charts dialog box as shown below. Click **OK**.

One-Way Tables & Charts

Data
Type of Data:
- ● Raw Categorical Data
- ○ Table of Frequencies

Raw Data Cell Range: A1:A100
☐ First cell contains label

Output Options
Title:

☐ Bar Chart
☐ Pie Chart
☐ Pareto Diagram

| Help | OK | Cancel |

10.   The output from Exercise 5B is displayed below. Because the numbers were generated randomly, it unlikely that your numbers will be exactly the same.

	A	B	C
1	One-Way Summary Table		
2			
3	Count of Variable		
4	Variable ▾	Total	
5	1	10	
6	2	6	
7	3	8	
8	4	14	
9	5	10	
10	6	9	
11	7	13	
12	8	7	
13	9	7	
14	10	5	
15	11	11	
16	Grand Total	100	

Discrete Probability Distributions

Section 4.2 Binomial Distributions

▶ Example 4 (pg. 210) | Finding a Binomial Probability Using Technology

1. In Try It Yourself 4, you are asked to find the probability that exactly 178 people will use more than one topping on their hot dogs. Open a new Excel worksheet and click in cell **A1** to place the output there.

2. Click **Formulas** and select **Insert Function**.

3. Select the **Statistical** category. Select the **BINOMDIST** function. Click **OK**.

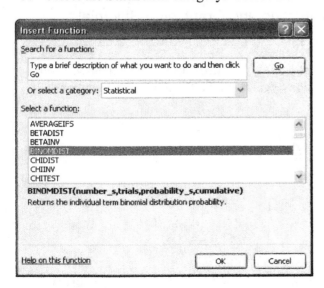

4. Complete the BINOMDIST dialog box as shown below. Click **OK**.

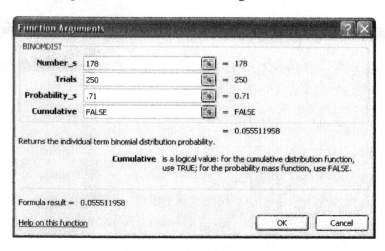

The BINOMDIST function returns a result of 0.055512.

	A
1	0.055512

► Example 7 (pg. 213)	Constructing and Graphing a Binomial Distribution

1. Open a new, blank Excel worksheet and enter the information shown below. You will be using the BINOMDIST function to calculate binomial probabilities for 0, 1, 2, 3, 4, 5, and 6 households.

	A	B	C
1	Households	Relative frequency	
2	0		
3	1		
4	2		
5	3		
6	4		
7	5		
8	6		

2. Click in cell **B2** below "Relative frequency." Click **Formulas** and select **Insert Function**.

3.   Under Function category, select **Statistical**. Under Function name, select
     BINOMDIST. Click **OK**.

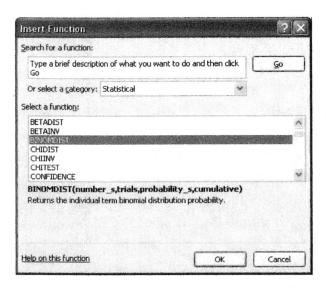

4.   Complete the BINOMDIST dialog box as shown below. Click **OK**.

You are entering a relative cell address without dollar signs (i.e., A2) in the
Number_s field because you will be copying the contents of cell B2 to cells B3
through B8. You want the column A cell address to change from A2 to A3, A4, A5,
..., A8 when the formula is copied from cell B2 to cells B3 through B8.

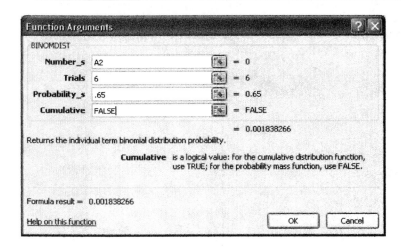

5.   Copy the contents of cell B2 to cells B3 through B8.

	A	B	C
1	Households	Relative frequency	
2	0	0.001838	
3	1	0.020484	
4	2	0.095102	
5	3	0.235491	
6	4	0.328005	
7	5	0.243661	
8	6	0.075419	

6.   Highlight cells **A1:B8**. Then click **Insert**, select a **Column** chart, and select the leftmost diagram in the 2-D Column row.

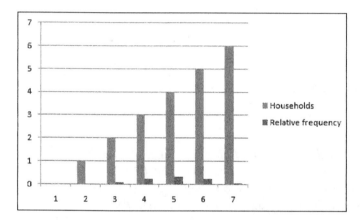

7.   Click **Design** near the top of the screen. and select **Select Data** in the Data group. In the Select Data Source dialog box, select **Households** by clicking on it. Then click the **Remove** button. Click **OK**.

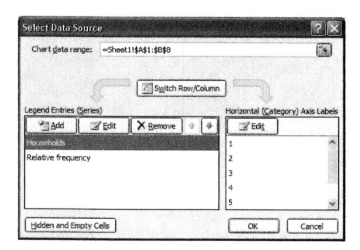

8. Click on the **Relative frequency** title in the chart and change it to **Subscribing to Cable TV**.

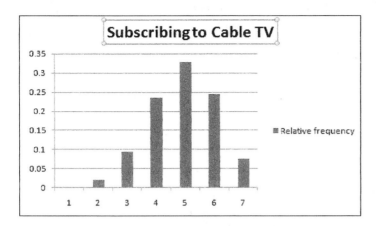

9. At the top of the screen, in the Chart Layouts group, select **Layout 8** so that you can add a Y-axis title and an X-axis title.   Change the Y-axis title to **Relative Frequency** and the X-axis title to **Households**.

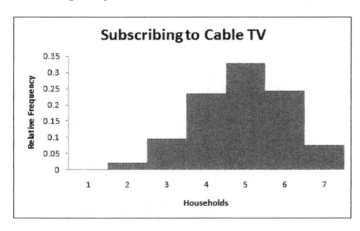

10. The X-axis number scale should be 0 to 6. These are the numbers in column A of the worksheet. To make this change, **right click** on any number in the X-axis in the chart. I clicked on 3. Click on **Select Data** in the menu that appears.

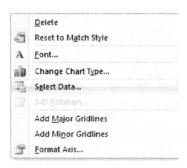

11. Click on **Edit** below Horizontal (Category) Axis Labels.   Click and drag over A2:A8 in the worksheet to enter that range in the Axis Labels dialog box. Click **OK**. Click **OK** in the Select Data Source dialog box.

The completed chart is shown below.

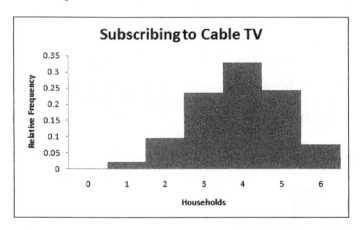

Section 4.3 More Discrete Probability Distributions

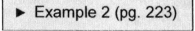 **► Example 2 (pg. 223)**   Finding a Poisson Probability

You will be using the POISSON function to find the probability that 4 accidents will occur at a certain intersection when the mean number of accidents per month at this intersection is 3.

1. Open a new Excel worksheet and click in cell **A1** to place the output there.

2. At the top of the screen, click **Formulas** and select **Insert Function**.

3.   Select the **Statistical** category. Select the **POISSON** function. Click **OK**.

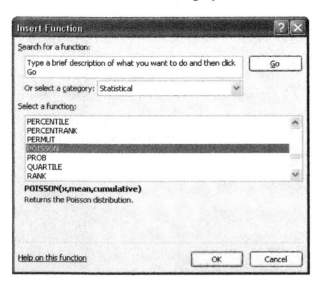

4.   Complete the POISSON dialog box as shown at the top of the next page. Click **OK**.

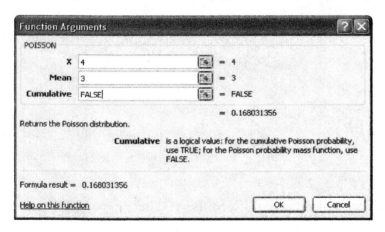

The POISSON function returns a result of 0.168031.

Technology

▶ Exercise 1 (pg. 237)	Creating a Poisson Distribution with μ = 4 for x = 0 to 20

1. Open a new Excel worksheet and enter the numbers 0 through 20 in column A as shown at the top of the next page. Then click in cell **B1** of the worksheet to place the output from the POISSON function there.

	A	B
1	0	
2	1	
3	2	
4	3	
5	4	
6	5	
7	6	
8	7	
9	8	
10	9	
11	10	
12	11	
13	12	
14	13	
15	14	
16	15	
17	16	
18	17	
19	18	
20	19	
21	20	

2. At the top of the screen, click **Formulas** and select **Insert Function**.

3.   Select the **Statistical** category. Select the **POISSON** function. Click **OK**.

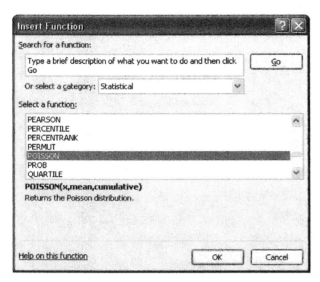

4.   Complete the POISSON dialog box as shown below. Click **OK**.

You are entering a relative cell address without dollar signs (i.e., A1) in the X field because you will be copying the contents of cell B1 to cells B2 through B11. You want the column A cell address to change from A1 to A2, A3, A4, ..., A21 when the formula is copied from cell B1 to cells B2 through B21.

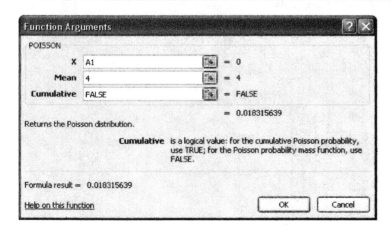

5. Copy the contents of cell B1 to cells B2 through B21.

	A	B
1	0	0.018316
2	1	0.073263
3	2	0.146525
4	3	0.195367
5	4	0.195367
6	5	0.156293
7	6	0.104196
8	7	0.05954
9	8	0.02977
10	9	0.013231
11	10	0.005292
12	11	0.001925
13	12	0.000642
14	13	0.000197
15	14	5.64E-05
16	15	1.5E-05
17	16	3.76E-06
18	17	8.85E-07
19	18	1.97E-07
20	19	4.14E-08
21	20	8.28E-09

6. Construct a histogram of the Poisson distribution. First highlight cells **A1:B21**.
    Then, at the top of the screen, click **Insert** and select a **Column** chart.

7. Select the leftmost diagram in the 2-D row.

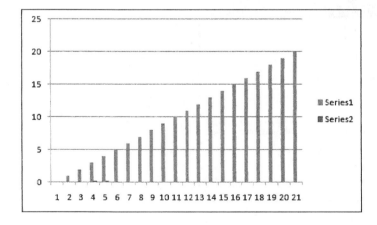

8. In the Chart Layouts section near the top of the screen, select **Layout 8** so that you can add a chart title, Y-axis title, and X-axis title.   Click on each title and change the chart title to **Customers Arriving at the Check-out Counter,** change the Y-axis title to **Probability**, change the X-axis title to **Number of Arrivals per Minute**.

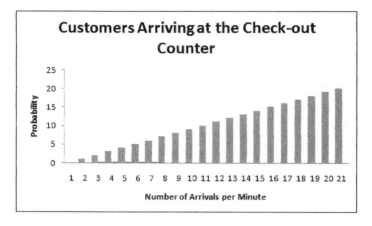

9.   Correct the number scales displayed on the Y-axis and the X-axis. **Right click** on one of the vertical bars. Then click on **Select Data** in the shortcut menu that appears.

10.   Click on **Series1.** Then click the **Remove** button. Click **OK**.

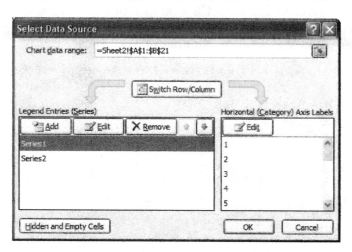

Your histogram should look similar to the one shown below.

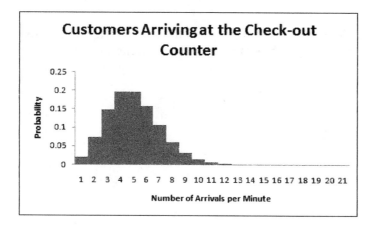

> ▶ **Exercise 3 (pg. 237)**   Generating a List of 20 Random Numbers
> with a Poisson Distribution for μ = 4

1. Open a new, blank Excel worksheet. At the top of the screen, click **Data** and select
    Data Analysis. Select **Random Number Generation** and click **OK**.

---

*If Data Analysis does not appear as a choice in the Data ribbon, you will need to
load the Microsoft Excel Analysis ToolPak add-in. Follow the procedure in Section
GS 8.1 before continuing.*

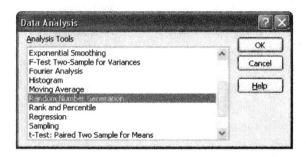

2. Complete the Random Number Generation dialog box as shown below. The
    Number of Variables indicates the number of columns of values that you want in
    the output table. **Number of Random Numbers** to be generated is 20. Select the
    Poisson distribution. **Lambda** is the expected value (μ), which is equal to 4. The
    output will be placed in the current worksheet with A1 as the left topmost cell.
    Click **OK**.

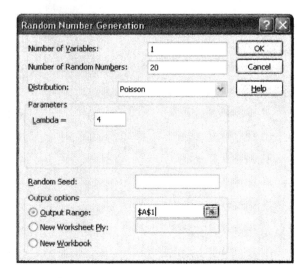

The 20 numbers generated for this example are shown at the top of the next page. Because the numbers were generated randomly, it is not likely that your numbers will be exactly the same.

	A
1	7
2	0
3	4
4	5
5	3
6	3
7	1
8	3
9	4
10	6
11	1
12	5
13	2
14	3
15	5
16	4
17	1
18	2
19	3
20	7

Normal Probability Distributions

Section 5.1 Introduction to Normal Distributions and the Standard Normal Distribution

► Example 4 (pg. 246) Finding Area Under the Standard Normal Curve

1. Open a new Excel worksheet and click in cell **A1** to place the output there.

2. At the top of the screen, click **Formulas** and select **Insert Function**.

3. Select the **Statistical** category. Select the **NORMSDIST** function. Click **OK**.

Be sure to select NORMSDIST and not NORMDIST.

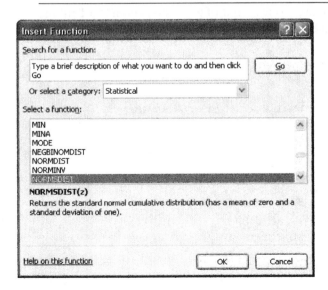

4. Complete the NORMSDIST dialog box as shown below. Click **OK**.

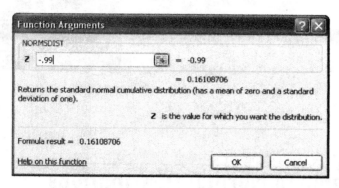

The output is displayed in cell A1 of the worksheet. The area under the standard normal curve to the left of $z = -0.99$ is 0.161087.

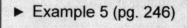

▶ Example 5 (pg. 246)   **Finding Area Under the Standard Normal Curve**

1. Open a new Excel worksheet. The area to the right of $z = 1.06$ is equal to 1 minus the area to the left of $z = 1.06$. In cell A1, enter **=1-** as shown below.

2. At the top of the screen, click **Formulas** and select **Insert Function**.

3. Select the **Statistical** category. Select the **NORMSDIST** function. Click **OK**.

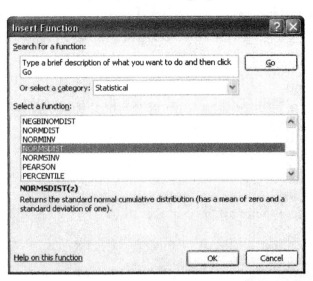

4. Complete the NORMSDIST dialog box as shown below. Click **OK**.

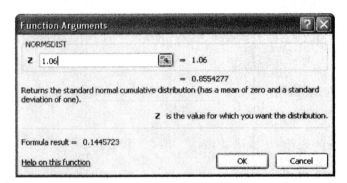

The output is displayed in cell A1 of the worksheet. The area under the standard normal curve to the right of z = 1.06 is 0.144572. In the formula bar, you should see f_x =1-NORMSDIST(1.06).

Section 5.2 Normal Distributions: Finding Probabilities

| ► Example 3 (pg. 255) | Using Technology to Find Normal Probabilities |

1.   Open a new Excel worksheet and click in cell **A1** to place the output there.

2.   At the top of the screen, click **Formulas** and select **Insert Function**.

3.   Select the **Statistical** category.   Select the **NORMDIST** function.   Click **OK**.

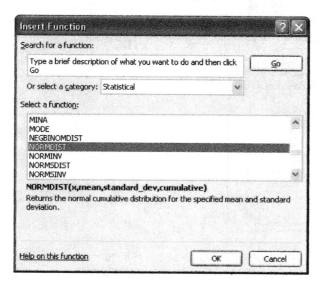

4.   Complete the NORMDIST dialog box as shown below. Click **OK**.

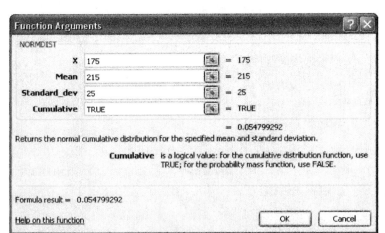

The output is displayed in cell A1 of the worksheet. The probability is equal to 0.054799.

5.3 Normal Distributions: Finding Values

► Example 2 (pg. 262)	Finding a z-Score Given a Percentile

1.   We will start with finding the z-score that corresponds to P_5. Open a new Excel worksheet and click in cell **A1** to place the output there.

2.   At the top of the screen, click **Formulas** and select **Insert Function**.

3.   Select the Statistical category. Select the **NORMSINV** function. Click **OK**.

Be sure to select NORMSINV and not NORMINV.

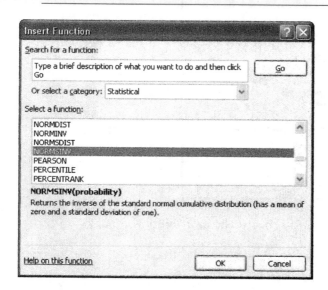

4.   Complete the NORMSINV dialog box as shown below. Click **OK**.

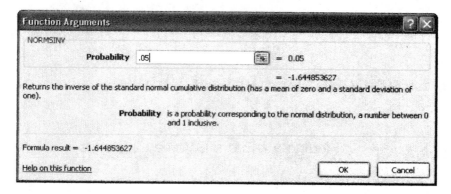

5.   The function returns a z-score of -1.64485. To find the z-scores for P50 and P90, repeat these steps using probabilities of .50 and .90, respectively.

▶ Example 4 (pg. 264)   Finding a Specific Data Value

1. You will find the lowest civil service exam score you can earn and still be eligible for employment. Open a new Excel worksheet and click in cell **A1** to place the output there.

2. At the top of the screen, click **Formulas** and select **Insert Function**.

3. Select the **Statistical** category. Select the **NORMINV** function. Click **OK**.

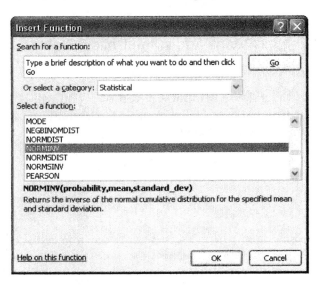

4. Complete the NORMINV dialog box as shown below. Click **OK**.

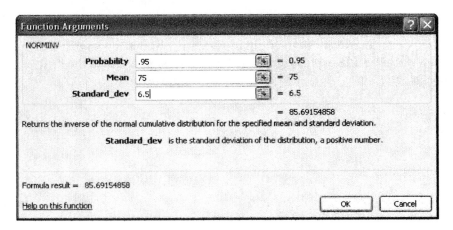

5.   The NORMINV function returns a value of 85.69155.

Technology

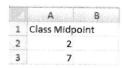

► Exercise 1 (pg. 303)   Finding the Mean Age in the United States

1.   Open a new, blank Excel worksheet. Enter **Class Midpoint** in cell A1. You will be entering the numbers displayed in the table on page 303 of your text. Begin by entering the first two midpoints, **2** and **7**, as shown below.

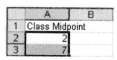

2.   You will now fill column A with a series. Click in cell A2 and drag down to cell A3 so that both cells are highlighted.

3.   Move the mouse pointer in cell A3 to the right lower corner of the cell so that the white plus sign turns into a black plus sign. The black plus sign is called the "fill handle." Click the left mouse button and drag the fill handle down to cell A21.

	A	B
1	Class Midpoint	
2	2	
3	7	
4	12	
5	17	
6	22	
7	27	
8	32	
9	37	
10	42	
11	47	
12	52	
13	57	
14	62	
15	67	
16	72	
17	77	
18	82	
19	87	
20	92	
21	97	

4. Enter **Relative Frequency** in cell **B1**. Then enter the proportion equivalents of the percentages in column B as shown below.

	A	B	C
1	Class Midpoint	Relative Frequency	
2	2	0.068	
3	7	0.066	
4	12	0.07	
5	17	0.071	
6	22	0.071	
7	27	0.068	
8	32	0.068	
9	37	0.071	
10	42	0.077	
11	47	0.076	
12	52	0.067	
13	57	0.059	
14	62	0.044	
15	67	0.034	
16	72	0.029	
17	77	0.025	
18	82	0.019	
19	87	0.011	
20	92	0.005	
21	97	0.001	

5. Click in cell **C2** and enter a formula to multiply the midpoint by the relative frequency. The formula is =A2*B2. Press [**Enter**].

	A	B	C
1	Class Midpoint	Relative Frequency	
2	2	0.068	=A2*B2
3	7	0.066	

6. Copy the contents of cell C2 to cells C3 through C21.

	A	B	C
1	Class Midpoint	Relative Frequency	
2	2	0.068	0.136
3	7	0.066	0.462
4	12	0.07	0.84
5	17	0.071	1.207
6	22	0.071	1.562
7	27	0.068	1.836
8	32	0.068	2.176
9	37	0.071	2.627
10	42	0.077	3.234
11	47	0.076	3.572
12	52	0.067	3.484
13	57	0.059	3.363
14	62	0.044	2.728
15	67	0.034	2.278
16	72	0.029	2.088
17	77	0.025	1.925
18	82	0.019	1.558
19	87	0.011	0.957
20	92	0.005	0.46
21	97	0.001	0.097

7. The weighted mean is equal to the sum of the products in column C. (Refer to Section 2.3 in your text.) Click in cell **C22** of the worksheet to place the sum there. Click **Formulas** near the top of the screen and select **AutoSum**. The range of numbers to be included in the sum is displayed in cell C22. You should see =SUM(C2:C21). Make any necessary corrections. Then press [**Enter**].

21	97	0.001	0.097
22			=SUM(C2:C21)

8. The mean, displayed in cell C22, is 36.59 years. Save this worksheet so that you can use it again for Exercise 5, page 303.

20	92	0.005	0.46
21	97	0.001	0.097
22			36.59

▶ Exercise 2 (pg. 303) | Finding the Mean of the Set of Sample Means

1. Open worksheet "Tech5" in the Chapter 5 folder.

2. Click in cell **A38** at the bottom of the column of numbers to place the mean in that cell.

36	31.27
37	35.80
38	

3. At the top of the screen, click **Formulas** and select **Insert Function**.

4. Select the **Statistical** category. Select the **AVERAGE** function. Click **OK**.

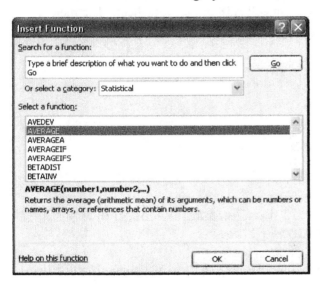

5. The range that will be included in the average is shown in the Number 1 window. Check to be sure that it is accurate. It should read **A2:A37**. Make any necessary corrections. Then click **OK**.

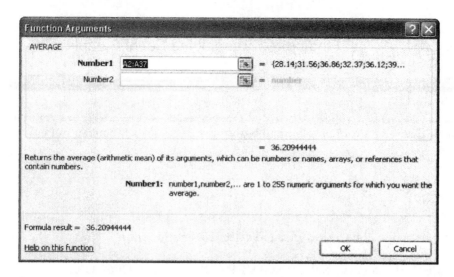

The mean of the sample means is displayed in cell A38 of the worksheet. The mean is equal to 36.21.

36	31.27
37	35.80
38	36.21

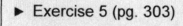

▶ Exercise 5 (pg. 303)   Finding the Standard Deviation of Ages in the United States

1. Open the worksheet that you prepared for Exercise 1, page 303. The first few lines are shown below.

	A	B	C
1	Class Midpoint	Relative Frequency	
2	2	0.068	0.136
3	7	0.066	0.462
4	12	0.07	0.84
5	17	0.071	1.207

2. To find the standard deviation of the population of ages, you will first calculate squared deviation scores. Click in cell **D1** and enter the label, **Sqd Dev Score**.

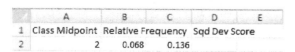

	A	B	C	D	E
1	Class Midpoint	Relative Frequency		Sqd Dev Score	
2	2	0.068	0.136		

3. Click in cell **D2** and enter the formula =**(A2-36.59)^2** and press [**Enter**].

	A	B	C	D	E
1	Class Midpoint	Relative Frequency	Sqd Dev Score		
2	2	0.068	0.136	=(A2-36.59)^2	

4. Copy the formula in cell D2 to cells D3 through D21.

	A	B	C	D	E
1	Class Midpoint	Relative Frequency	Sqd Dev Score		
2	2	0.068	0.136	1196.468	
3	7	0.066	0.462	875.5681	
4	12	0.07	0.84	604.6681	
5	17	0.071	1.207	383.7681	
6	22	0.071	1.562	212.8681	
7	27	0.068	1.836	91.9681	
8	32	0.068	2.176	21.0681	
9	37	0.071	2.627	0.1681	
10	42	0.077	3.234	29.2681	
11	47	0.076	3.572	108.3681	
12	52	0.067	3.484	237.4681	
13	57	0.059	3.363	416.5681	
14	62	0.044	2.728	645.6681	
15	67	0.034	2.278	924.7681	
16	72	0.029	2.088	1253.868	
17	77	0.025	1.925	1632.968	
18	82	0.019	1.558	2062.068	
19	87	0.011	0.957	2541.168	
20	92	0.005	0.46	3070.268	
21	97	0.001	0.097	3649.368	

5. Each of the squared deviations will be weighted by its relative frequency. Click in cell **E2**. Enter the formula =**D2*B2** and press [**Enter**].

	A	B	C	D	E
1	Class Midpoint	Relative Frequency	Sqd Dev Score		
2	2	0.068	0.136	1196.468	=D2*B2

6. Copy the formula in cell E2 to cells E3 through E21.

	A	B	C	D	E
1	Class Midpoint	Relative Frequency		Sqd Dev	Score
2	2	0.068	0.136	1196.468	81.35983
3	7	0.066	0.462	875.5681	57.78749
4	12	0.07	0.84	604.6681	42.32677
5	17	0.071	1.207	383.7681	27.24754
6	22	0.071	1.562	212.8681	15.11364
7	27	0.068	1.836	91.9681	6.253831
8	32	0.068	2.176	21.0681	1.432631
9	37	0.071	2.627	0.1681	0.011935
10	42	0.077	3.234	29.2681	2.253644
11	47	0.076	3.572	108.3681	8.235976
12	52	0.067	3.484	237.4681	15.91036
13	57	0.059	3.363	416.5681	24.57752
14	62	0.044	2.728	645.6681	28.4094
15	67	0.034	2.278	924.7681	31.44212
16	72	0.029	2.088	1253.868	36.36217
17	77	0.025	1.925	1632.968	40.8242
18	82	0.019	1.558	2062.068	39.17929
19	87	0.011	0.957	2541.168	27.95285
20	92	0.005	0.46	3070.268	15.35134
21	97	0.001	0.097	3649.368	3.649368

7. You are working with a relative frequency distribution of midpoints. For this type of distribution, the variance is equal to the sum of the squared deviation scores. Click in cell **E22** to place the sum of the squared deviation scores there. Click **Formulas** near the top of the screen and select **AutoSum.** The range of numbers to be included in the sum is displayed in cell E22. You should see =SUM(E2:E21). Make any necessary corrections. Then press [**Enter**].

20	92	0.005	0.46	3070.268	15.35134
21	97	0.001	0.097	3649.368	3.649368
22			36.59		=SUM(E2:E21)

8. Click in cell **E23** to place the standard deviation there. The standard deviation is the square root of the variance. In cell E23, enter the formula =**sqrt(E22)** and press [**Enter**].

21	97	0.001	0.097	3649.368	3.649368
22			36.59		505.6819
23					=sqrt(E22)

9. The standard deviation of ages in the United States is approximately 22.49. For future reference purposes, you may want to add labels for the mean, variance, and standard deviation as shown below.

20		92	0.005		0.46	3070.268	15.35134
21		97	0.001		0.097	3649.368	3.649368
22			Mean =		36.59	Var =	505.6819
23						St Dev =	22.48737

◀

▶ **Exercise 6 (pg. 303)**　　Finding the Standard Deviation of the Set of Sample Means

1. Open worksheet "Tech5" in the Chapter 5 folder. This is the same data set that you used for Exercise 2, page 303. If you placed the mean (36.21) in cell A38 for Exercise 2, page 303, you should delete it now.

	A
1	Mean ages
2	28.14
3	31.56
4	36.86

2. At the top of the screen, select **Data** and **Data Analysis**.

If Data Analysis does not appear as a choice in the Data ribbon, you will need to load the Microsoft Excel Analysis ToolPak add-in. Follow the procedure in Section GS 8.1 before continuing.

---

3. Select **Descriptive Statistics** and click **OK**.

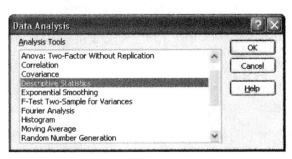

4. Complete the Descriptive Statistics dialog box as shown below. Be sure to select **Summary statistics** near the bottom of the dialog box. Click **OK**.

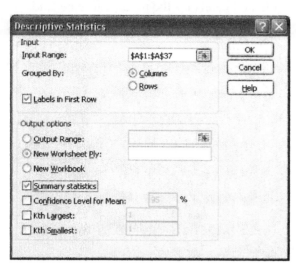

You will want to make column A wider so that you can read all the labels in the output table. The standard deviation is equal to 3.551804.

	A	B
1	Mean ages	
2		
3	Mean	36.20944
4	Standard Error	0.591967
5	Median	36.155
6	Mode	#N/A
7	Standard Deviation	3.551804
8	Sample Variance	12.61531
9	Kurtosis	0.343186
10	Skewness	0.207283
11	Range	16.58
12	Minimum	28.14
13	Maximum	44.72
14	Sum	1303.54
15	Count	36

Confidence Intervals

Section 6.1 Confidence Intervals for the Mean (Large Samples)

▶ Example 4 (pg. 314)	Constructing a Confidence Interval Using Technology

If the PHStat2 add-in has not been loaded, you will need to load it before continuing. Follow the instructions in Section GS 8.2.

1. Open the "Sentences" worksheet in the Chapter 6 folder.

2. At the top of the screen, select **Add-Ins** and **PHStat**. Select **Confidence Intervals** → **Estimate for the Mean, sigma known**.

3. Complete the Estimate for the Mean dialog box as shown at the top of the next page. Click **OK**.

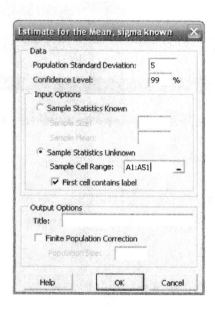

The output is displayed in a new worksheet. The lower limit of the confidence interval is 10.579 and the upper limit is 14.221.

	A	B
1	Confidence Interval Estimate for the Mean	
2		
3	Data	
4	Population Standard Deviation	5
5	Sample Mean	12.4
6	Sample Size	50
7	Confidence Level	99%
8		
9	Intermediate Calculations	
10	Standard Error of the Mean	0.707106781
11	Z Value	-2.5758293
12	Interval Half Width	1.821386368
13		
14	Confidence Interval	
15	Interval Lower Limit	10.57861363
16	Interval Upper Limit	14.22138637

▶ Example 5 (pg. 315)	Constructing a Confidence Interval, σ Known

If the PHStat2 add-in has not been loaded, you will need to load it before continuing. Follow the instructions in Section GS 8.2.

1. Open a new Excel worksheet. At the top of the screen, select **Add-Ins** and **PHStat**. Select **Confidence Intervals → Estimate for the Mean, sigma known**.

2. Complete the Estimate for the Mean dialog box as shown below. Click **OK**.

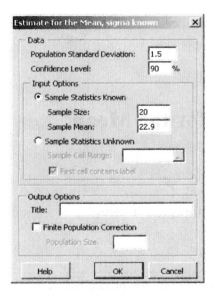

Your output should look the same as the output displayed at the top of the next page. The lower limit of the 90% confidence interval is 22.3 and the upper limit is 23.5.

	A	B
1	Confidence Interval Estimate for the Mean	
2		
3	Data	
4	Population Standard Deviation	1.5
5	Sample Mean	22.9
6	Sample Size	20
7	Confidence Level	90%
8		
9	Intermediate Calculations	
10	Standard Error of the Mean	0.335410197
11	Z Value	-1.64485363
12	Interval Half Width	0.551700678
13		
14	Confidence Interval	
15	Interval Lower Limit	22.34829932
16	Interval Upper Limit	23.45170068

◀

Section 6.2 Confidence Intervals for the Mean (Small Samples)

▶ Example 2 (pg. 327)   Constructing a Confidence Interval

If the PHStat2 add-in has not been loaded, you will need to load it before continuing. Follow the instructions in Section GS 8.2.

1. Open a new Excel worksheet. At the top of the screen, select **Add-Ins** and **PHStat**. Select **Confidence Intervals → Estimate for the Mean, sigma unknown**.

2. Complete the Estimate for the Mean dialog box as shown at the top of the next page. Click **OK**.

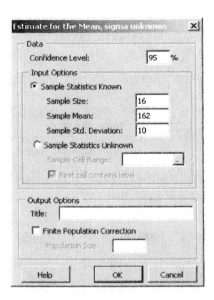

The output is displayed in a new worksheet. The lower limit of the 95% confidence interval is 156.7 and the upper limit is 167.3.

	A	B	C
1	**Confidence Interval Estimate for the Mean**		
2			
3	**Data**		
4	**Sample Standard Deviation**	10	
5	**Sample Mean**	162	
6	**Sample Size**	16	
7	**Confidence Level**	95%	
8			
9	**Intermediate Calculations**		
10	Standard Error of the Mean	2.5	
11	Degrees of Freedom	15	
12	*t* Value	2.131449536	
13	Interval Half Width	5.328623839	
14			
15	**Confidence Interval**		
16	**Interval Lower Limit**	156.67	
17	**Interval Upper Limit**	167.33	

Section 6.3 Confidence Intervals for Population Proportions

▶ Example 2 (pg. 336)   Constructing a Confidence Interval for p

The problem asks you to construct a 95% confidence interval for the proportion of adults in the United States who say that their favorite sport to watch is football.

If the PHStat2 add-in has not been loaded, you will need to load it before continuing. Follow the instructions in Section GS 8.2.

1. Open a new Excel worksheet. At the top of the screen, select **Add-Ins** and **PHStat**. Select **Confidence Intervals → Estimate for the Proportion**.

2. Complete the Estimate for the Proportion dialog box as shown below. Click **OK**.

Your output should look like the output displayed at the top of the next page. The lower limit of the 95% confidence interval is 0.265 and the upper limit is 0.316.

	A	B
1	**Confidence Interval Estimate for the Mean**	
2		
3	**Data**	
4	**Sample Size**	1219
5	**Number of Successes**	354
6	**Confidence Level**	95%
7		
8	Intermediate Calculations	
9	Sample Proportion	0.290401969
10	Z Value	-1.95996398
11	Standard Error of the Proportion	0.013001819
12	Interval Half Width	0.025483098
13		
14	**Confidence Interval**	
15	**Interval Lower Limit**	0.264918871
16	**Interval Upper Limit**	0.315885066

◀

6.4 Confidence Intervals for Variation and Standard Deviation

▶ Example 2 (pg. 347) | Constructing a Confidence Interval

If the PHStat2 add-in has not been loaded, you will need to load it before continuing. Follow the instructions in Section GS 8.2.

3. Open a new Excel worksheet. At the top of the screen, select **Add-Ins** and **PHStat**. Select **Confidence Intervals → Estimate for the Population Variance**.

4. Complete the Estimate for the Population Variance dialog box as shown at the top of the next page. Click **OK**.

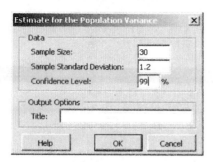

Your output should look the same as the output displayed below. The left and right endpoints of the confidence interval for σ^2 are 0.80 and 3.18, respectively.   The left and right endpoints of the confidence interval for σ are 0.89 and 1.78, respectively.

	A	B	C	D	E
1	**Confidence Interval Estimate for the Population Variance**				
2					
3	**Data**				
4	**Sample Size**	30			
5	**Sample Standard Deviation**	1.2			
6	**Confidence Level**	99%			
7					
8	Intermediate Calculations				
9	Degrees of Freedom	29			
10	Sum of Squares	41.76			
11	Single Tail Area	0.005			
12	Lower Chi-Square Value	13.12115			
13	Upper Chi-Square Value	52.33562			
14					
15	Results				
16	**Interval Lower Limit for Variance**	0.797927			
17	**Interval Upper Limit for Variance**	3.182648			
18					
19	**Interval Lower Limit for Standard Deviation**	0.893268			
20	**Interval Upper Limit for Standard Deviation**	1.783998			
21					
22	*Assumption:*				
23	**Population from which sample was drawn has an approximate normal distribution.**				

Technology

▶ Exercise 1 (pg. 359)   Finding a 95% Confidence Interval for p

If the PHStat2 add-in has not been loaded, you will need to load it before continuing. Follow the instructions in Section GS 8.2.

1. Open a new Excel worksheet. At the top of the screen, select **Add-Ins** and **PHStat**.
    Select **Confidence Intervals → Estimate for the Proportion**.

2. Interviews were completed with 1,010 adults. Thirteen percent or 131 adults named
    George W. Bush. Complete the Estimate for the Proportion dialog box as shown
    below. Click **OK**.

Your output should look the same as the output displayed below. The lower limit of the
95% confidence interval is 0.109 and the upper limit is 0.150.

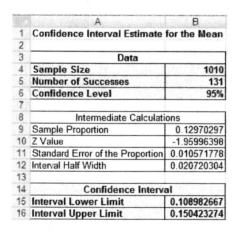

	A	B
1	**Confidence Interval Estimate for the Mean**	
2		
3	**Data**	
4	**Sample Size**	1010
5	**Number of Successes**	131
6	**Confidence Level**	95%
7		
8	Intermediate Calculations	
9	Sample Proportion	0.12970297
10	Z Value	-1.95996398
11	Standard Error of the Proportion	0.010571778
12	Interval Half Width	0.020720304
13		
14	Confidence Interval	
15	**Interval Lower Limit**	0.108982667
16	**Interval Upper Limit**	0.150423274

▶ **Exercise 3 (pg. 359)**   Finding a 95% Confidence Interval for p

The problem asks you to find a 95% confidence interval for the proportion of the
population that would have chosen Oprah Winfrey.

> *If the PHStat2 add-in has not been loaded, you will need to load it before continuing. Follow the instructions in Section GS 8.2.*

1. Open a new Excel worksheet. At the top of the screen, select **Add-Ins** and **PHStat**. Select **Confidence Intervals** → **Estimate for the Proportion**.

2. Nine percent of the sample size, 1010, is 91. Use this number to complete the Estimate for the Proportion dialog box as shown below. Click **OK**.

Your output should look the same as the output displayed below. The lower limit of the 95% confidence interval is 0.072 and the upper limit is 0.108.

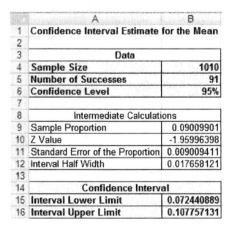

	A	B
1	**Confidence Interval Estimate for the Mean**	
2		
3	**Data**	
4	**Sample Size**	1010
5	**Number of Successes**	91
6	**Confidence Level**	95%
7		
8	**Intermediate Calculations**	
9	Sample Proportion	0.09009901
10	Z Value	-1.95996398
11	Standard Error of the Proportion	0.009009411
12	Interval Half Width	0.017658121
13		
14	**Confidence Interval**	
15	**Interval Lower Limit**	0.072440889
16	**Interval Upper Limit**	0.107757131

▶ Exercise 4 (pg. 359) | Simulating a Most Admired Poll

1. Open a new, blank Excel worksheet. Enter the labels shown below for displaying the output of five simulations.

2. At the top of the screen, select **Data** and **Data Analysis**. Select **Random Number Generation**. Click **OK**.

If Data Analysis does not appear as a choice in the Data ribbon, you will need to load the Microsoft Excel Analysis ToolPak add-in. Follow the procedure in Section GS 8.1 before continuing.

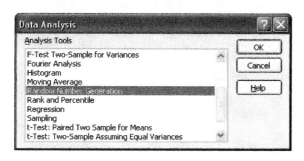

3.   Complete the Random Number Generation dialog box as shown below. Click **OK**.

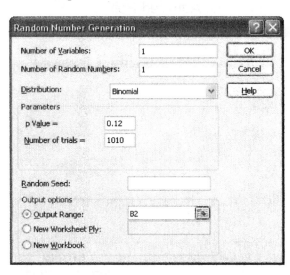

For Time 1, the number of persons (out of 1010) selecting Oprah Winfrey is 122.
Because this number was generated randomly, it is unlikely that your number is
exactly the same.

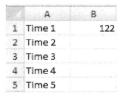

	A	B
1	Time 1	122
2	Time 2	
3	Time 3	
4	Time 4	
5	Time 5	

4.   Begin Time 2 the same way that you did Time 1. At the top of the screen, select
     Data and **Data Analysis**. Select **Random Number Generation**. Click **OK**.

5.   Complete the Random Number Generation dialog box in the same way as you did
     for Time 1 except change the location of the output to cell B2 as shown at the top
     of the next page. Click **OK**.

6. Repeat this procedure until you have carried out the simulation five times. The output I generated is displayed below.

	A	B
1	Time 1	122
2	Time 2	119
3	Time 3	137
4	Time 4	121
5	Time 5	121

Hypothesis Testing with One Sample

Section 7.2 Hypothesis Testing for the Mean (Large Samples)

▶ Example 6 (pg. 384)	Using a Technology Tool to Find a P-value

If the PHStat add-in has not been loaded, you will need to load it before continuing. Follow the instructions in Section GS 8.2.

1. First open a new Excel worksheet. Then, at the top of the screen, select **Add-Ins** and **PHStat**. Select **One-Sample Tests→Z Test for the Mean, sigma known**.

2. Complete **the** Z Test for the Mean dialog box as shown at the top of the next page. Click **OK**.

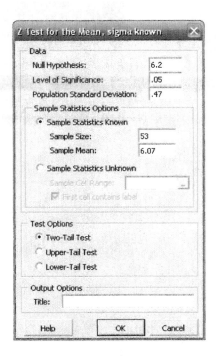

Your output should look like the output displayed below.

	A	B
1	Z Test of Hypothesis for the Mean	
2		
3	Data	
4	Null Hypothesis μ=	6.2
5	Level of Significance	0.05
6	Population Standard Deviation	0.47
7	Sample Size	53
8	Sample Mean	6.07
9		
10	Intermediate Calculations	
11	Standard Error of the Mean	0.064559465
12	Z Test Statistic	-2.013647416
13		
14	Two-Tail Test	
15	Lower Critical Value	-1.959963985
16	Upper Critical Value	1.959963985
17	p-Value	0.044046564
18	Reject the null hypothesis	

> ▶ Example 9 (pg. 387)

Testing μ with a Large Sample

If the PHStat add-in has not been loaded, you will need to load it before continuing. Follow the instructions in Section GS 8.2.

---

1. First open a new Excel worksheet. Then, at the top of the screen, select **Add-Ins** and **PHStat**. Select **One-Sample Tests→Z Test for the Mean, sigma known**.

2. Complete the Z Test for the Mean dialog box as shown below.   Be sure to select **Lower-Tail Test**. Click **OK**.

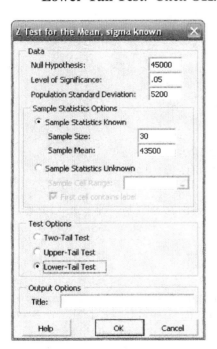

Your output should look like the output displayed below.

	A	B
1	Z Test of Hypothesis for the Mean	
2		
3	Data	
4	Null Hypothesis μ=	45000
5	Level of Significance	0.05
6	Population Standard Deviation	5200
7	Sample Size	30
8	Sample Mean	43500
9		
10	Intermediate Calculations	
11	Standard Error of the Mean	949.3857663
12	Z Test Statistic	-1.579968916
13		
14	Lower-Tail Test	
15	Lower Critical Value	-1.644853627
16	p-Value	0.057056993
17	Do not reject the null hypothesis	

◄

► Exercise 37 (pg. 392)   **Testing That the Mean Time to Quit Smoking Is 15 Years**

If the PHStat add-in has not been loaded, you will need to load it before continuing. Follow the instructions in Section GS 8.2.

---

1. Open the "Ex7_2-37" worksheet in the Chapter 7 folder.

2. Your textbook indicates that the standard deviation of a sample may be used in the z-test formula instead of a known population standard deviation if the sample size is 30 or greater. You will use Excel's Descriptive Statistics tool to obtain the sample standard deviation. At the top of the screen, select **Data** and **Data Analysis**. Select **Descriptive Statistics**. Click **OK**.

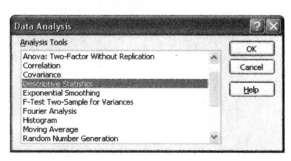

3. Complete the Descriptive Statistics dialog box as shown below. Be sure to select **Labels in first row** and **Summary statistics**. Click **OK**.

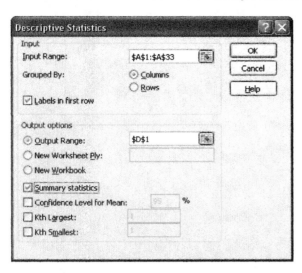

4. You will want to adjust the width of column D so that you can read all the labels. The standard deviation of the sample is 4.2876, the mean is 14.8344, and the count is 32. You will use this information when you carry out the statistical test. At the top of the screen, select **Add-Ins** and **PHStat**. Select **One-Sample Tests→Z Test for the Mean, sigma known**.

	A	B	C	D	E
1	Years			Years	
2	15.7				
3	13.2			Mean	14.83438
4	22.6			Standard Error	0.75795
5	13.0			Median	14.65
6	10.7			Mode	13.2
7	18.1			Standard Deviation	4.287612
8	14.7			Sample Variance	18.38362
9	7.0			Kurtosis	-0.57155
10	17.3			Skewness	0.042962
11	7.5			Range	15.6
12	21.8			Minimum	7
13	12.3			Maximum	22.6
14	19.8			Sum	474.7
15	13.8			Count	32

5. Complete the Z Test for the Mean dialog box as shown below. Click **OK**.

Your output should look like the output displayed below.

	A	B
1	Z Test of Hypothesis for the Mean	
2		
3	Data	
4	Null Hypothesis $\mu=$	15
5	Level of Significance	0.05
6	Population Standard Deviation	4.2876
7	Sample Size	32
8	Sample Mean	14.8344
9		
10	Intermediate Calculations	
11	Standard Error of the Mean	0.757947759
12	Z Test Statistic	-0.218484715
13		
14	Two-Tail Test	
15	Lower Critical Value	-1.959963985
16	Upper Critical Value	1.959963985
17	*p*-Value	0.827051466
18	Do not reject the null hypothesis	

Section 7.3 Hypothesis Testing for the Mean (Small Samples)

► Example 4 (pg. 400)	Testing μ with a Small Sample

If the PHStat add-in has not been loaded, you will need to load it before continuing. Follow the instructions in Section GS 8.2

1. First open a new Excel worksheet. Then, at the top of the screen, select **Add-Ins** and **PHStat**. Select **One-Sample Tests→t Test for the Mean, sigma unknown**.

2. Complete the t Test for the Mean dialog box as shown below. Click **OK**.

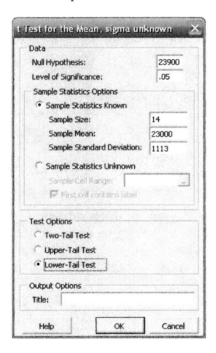

Your output should look like the output shown below.

	A	B	C	D	E
1	t Test for Hypothesis of the Mean				
2					
3	Data				
4	Null Hypothesis μ=	23900			
5	Level of Significance	0.05			
6	Sample Size	14			
7	Sample Mean	23000			
8	Sample Standard Deviation	1113			
9					
10	Intermediate Calculations				
11	Standard Error of the Mean	297.4617622			
12	Degrees of Freedom	13			
13	t Test Statistic	-3.025598965			
14					
15	Lower-Tail Test			Calculations Area	
16	Lower Critical Value	-1.770933383		For one-tailed tests:	
17	p-Value	0.004873492		TDIST value	0.004873
18	Reject the null hypothesis			1-TDIST value	0.995127

◀

▶ **Example 6 (pg. 402)** | **Using P-Values with a t-Test**

If the PHStat add-in has not been loaded, you will need to load it before continuing.
Follow the instructions in Section GS 8.2.

---

1. First open a new Excel worksheet. Then, at the top of the screen, select **Add-Ins** and **PHStat**. Select **One-Sample Tests→t Test for the Mean, sigma unknown**.

2. Complete the t Test for the Mean dialog box as shown at the top of the next page. Click **OK**.

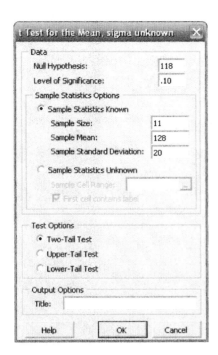

Your output should look like the output displayed below.

	A	B
1	t Test for Hypothesis of the Mean	
2		
3	Data	
4	Null Hypothesis μ=	118
5	Level of Significance	0.1
6	Sample Size	11
7	Sample Mean	128
8	Sample Standard Deviation	20
9		
10	Intermediate Calculations	
11	Standard Error of the Mean	6.030226892
12	Degrees of Freedom	10
13	t Test Statistic	1.658312395
14		
15	Two-Tail Test	
16	Lower Critical Value	-1.812461102
17	Upper Critical Value	1.812461102
18	p-Value	0.128245992
19	Do not reject the null hypothesis	

> ▶ Exercise 25 (pg. 404) | Testing the Claim That the Mean Recycled Waste Is More Than 1 Lb. Per Day

---

If the PHStat add-in has not been loaded, you will need to load it before continuing. Follow the instructions in Section GS 8.2.

---

1. First open a new Excel worksheet. Then, at the top of the screen, select **Add-Ins** and **PHStat**. Select **One-Sample Tests→t Test for the Mean, sigma unknown**.

2. Complete the t Test for the Mean dialog box as shown below. Be sure to select **Upper-Tail Test**. Click **OK**.

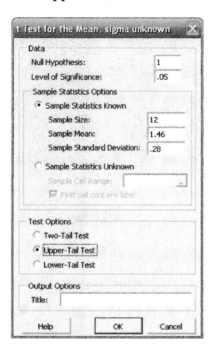

Your output should be the same as the output displayed below.

	A	B	C	D	E
1	t Test for Hypothesis of the Mean				
2					
3	Data				
4	Null Hypothesis μ=	1			
5	Level of Significance	0.05			
6	Sample Size	12			
7	Sample Mean	1.46			
8	Sample Standard Deviation	0.28			
9					
10	Intermediate Calculations				
11	Standard Error of the Mean	0.080829038			
12	Degrees of Freedom	11			
13	t Test Statistic	5.691024082			
14					
15	Upper-Tail Test			Calculations Area	
16	Upper Critical Value	1.795884814		For one-tailed tests:	
17	p-Value	7.00186E-05		TDIST value	7E-05
18	Reject the null hypothesis			1-TDIST value	0.99993

◀

▶ Exercise 30 (pg. 405)   **Testing the Claim That Teachers Spend a Mean of More Than $550**

If the PHStat add-in has not been loaded, you will need to load it before continuing. Follow the instructions in Section GS 8.2.

---

1. Open the "Ex7_3-30" worksheet in the Chapter 7 folder.

2. At the top of the screen, select **Add-Ins** and **PHStat**. Select **One-Sample Tests→t Test for the Mean, sigma unknown**.

3. Complete the t Test for the Mean dialog box as shown below. Click **OK**.

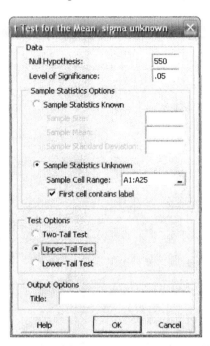

Your output should look like the output displayed below.

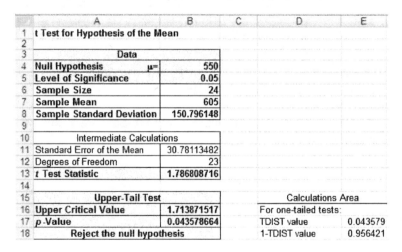

	A	B	C	D	E
1	t Test for Hypothesis of the Mean				
2					
3	Data				
4	Null Hypothesis μ=	550			
5	Level of Significance	0.05			
6	Sample Size	24			
7	Sample Mean	605			
8	Sample Standard Deviation	150.796148			
9					
10	Intermediate Calculations				
11	Standard Error of the Mean	30.78113482			
12	Degrees of Freedom	23			
13	t Test Statistic	1.786808716			
14					
15	Upper-Tail Test			Calculations Area	
16	Upper Critical Value	1.713871517		For one-tailed tests:	
17	p-Value	0.043578664		TDIST value	0.043579
18	Reject the null hypothesis			1-TDIST value	0.956421

Section 7.4 Hypothesis Testing for Proportions

► Example 1 (pg. 408) | Hypothesis Test for a Proportion

If the PHStat add-in has not been loaded, you will need to load it before continuing. Follow the instructions in Section GS 8.2.

1. First open a new Excel worksheet. Then, at the top of the screen, select **Add-Ins** and **PHStat**. Select **One-Sample Tests→Z Test for the Proportion**.

2. Complete the Z Test for the Proportion dialog box as shown at the top of the next page. Click **OK**.

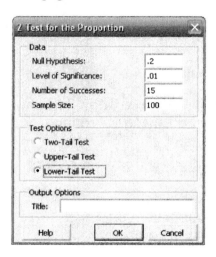

Your output should be the same as the output displayed below.

	A	B	C
1	Z Test of Hypothesis for the Proportion		
2			
3	Data		
4	Null Hypothesis p=	0.2	
5	Level of Significance	0.01	
6	Number of Successes	15	
7	Sample Size	100	
8			
9	Intermediate Calculations		
10	Sample Proportion	0.15	
11	Standard Error	0.04	
12	Z Test Statistic	-1.25	
13			
14	Lower-Tail Test		
15	Lower Critical Value	-2.326347874	
16	p-Value	0.105649774	
17	Do not reject the null hypothesis		

◄

► Exercise 11 (pg. 411)	Testing the Claim That More Than 30% of Consumers Stopped Buying a Product

If the PHStat add-in has not been loaded, you will need to load it before continuing. Follow the instructions in Section GS 8.2.

1. First open a new Excel worksheet. Then, at the top of the screen, select **Add-Ins** and **PHStat**. Select **One-Sample Tests→Z Test for the Proportion**.

2. Complete the Z Test for the Proportion dialog box as shown below. Click **OK**.

Note that 32% of 1050 is 336.

Your output should be the same as the output displayed below.

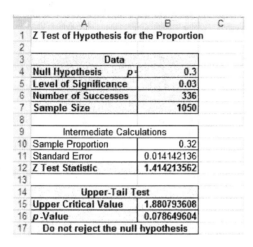

Section 7.5 Hypothesis Testing for Variance and Standard Deviation

---

▶ Exercise 26 (pg. 422)	Using the P-Value Method to Perform the Hypothesis Test for Exercise 22

If the PHStat add-in has not been loaded, you will need to load it before continuing. Follow the instructions in Section GS 8.2.

---

1. Open a new Excel worksheet. At the top of the screen, select **Add-Ins** and **PHStat**. Select **One-Sample Tests→Chi-Square Test for the Variance.**

2. Complete the Chi-Square Test for the Variance dialog box as shown below. Click **OK**.

Note that the null hypothesis value is the variance. For this problem, the variance is 30^2 or 900.

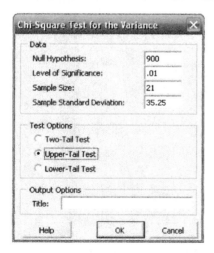

Your output should be the same as the output displayed at the top of the next page.

	A	B
1	Chi-Square Test of Variance	
2		
3	Data	
4	Null Hypothesis σ^2=	900
5	Level of Significance	0.01
6	Sample Size	21
7	Sample Standard Deviation	35.25
8		
9	Intermediate Calculations	
10	Degrees of Freedom	20
11	Half Area	0.005
12	Chi-Square Statistic	27.6125
13		
14	Upper-Tail Test	
15	Upper Critical Value	37.56623475
16	p-Value	0.118894324
17	Do not reject the null hypothesis	

◀

Technology

| ▶ Exercise 1 (pg. 433) | Testing the Claim That the Proportion Is Equal to 0.53 |

If the PHStat add-in has not been loaded, you will need to load it before continuing. Follow the instructions in Section GS 8.2.

1. First open a new Excel worksheet. Then, at the top of the screen, select **Add-Ins** and **PHStat**. Select **One-Sample Tests→Z Test for the Proportion**.

2. Complete the Z Test for the Proportion dialog box as shown at the top of the next page. Click **OK**.

Note that 0.2914 × 350 = 102. Note also that the Minitab illustration displays a 99.0% confidence interval. Therefore, the level of significance must be 0.01.

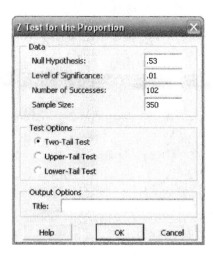

Your output should appear the same as the output shown below.

	A	B	C
1	Z Test of Hypothesis for the Proportion		
2			
3	Data		
4	Null Hypothesis p=	0.53	
5	Level of Significance	0.01	
6	Number of Successes	102	
7	Sample Size	350	
8			
9	Intermediate Calculations		
10	Sample Proportion	0.291428571	
11	Standard Error	0.026677974	
12	Z Test Statistic	-8.942636739	
13			
14	Two-Tail Test		
15	Lower Critical Value	-2.575829304	
16	Upper Critical value	2.575829304	
17	p-Value	0	
18	Reject the null hypothesis		

Hypothesis Testing with Two Samples

CHAPTER

8

Section 8.1 Testing the Difference Between Means (Large Independent Samples)

▶ Exercise 29 (pg. 447)
Testing the Claim That Children Ages 6-17 Watched TV More in 1981 Than Today

1. Open worksheet "Ex8_1-29" in the Chapter 8 folder.

2. Your textbook indicates that the variance of a sample may be used in the z-test formula instead of a known population variance if the sample size is sufficiently large. You will use Excel's Descriptive Statistics tool to obtain the sample variance. At the top of the screen, select **Data** and **Data Analysis**. Select **Descriptive Statistics**. Click **OK**.

If Data Analysis does not appear as a choice in the Data ribbon, you will need to load the Microsoft Excel ToolPak add-in. Follow the procedure in Section GS 8.1 before continuing.

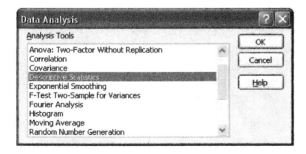

5. At the top of the screen, select **Data** and **Data Analysis**. Select **z-Test: Two Sample for Means** and click **OK**.

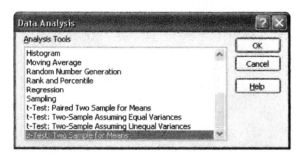

6. Complete the z-Test: Two Sample for Means dialog box as shown below. Click **OK**.

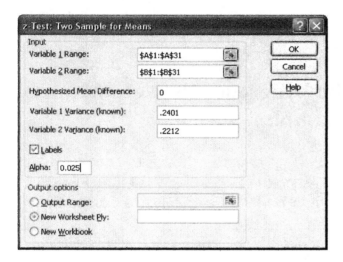

The output is displayed in a new worksheet. You will want to adjust the width of column A so that you can read the labels. Your output should look similar to the output displayed below.

	A	B	C
1	z-Test: Two Sample for Means		
2			
3		Time A	Time B
4	Mean	2.13	1.756667
5	Known Variance	0.2401	0.2212
6	Observations	30	30
7	Hypothesized Mean Difference	0	
8	z	3.010687	
9	P(Z<=z) one-tail	0.001303	
10	z Critical one-tail	1.959964	
11	P(Z<=z) two-tail	0.002607	
12	z Critical two-tail	2.241403	

◀

Section 8.2 Testing the Difference Between Means (Small Independent Samples)

▶ **Exercise 15 (pg. 457)**   Testing the Claim That the Mean Footwell Intrusion for Small & Midsize Cars Is Equal

If the PHStat add-in has not been loaded, you will need to load it before continuing. Follow the instructions in Section GS 8.2.

1. First open a new Excel worksheet. Then, at the top of the screen, select **Add-Ins** and **PHStat**. Select **Two-Sample Tests→t Test for Differences in Two Means**.

2. Complete the t Test for Differences in Two Means dialog box as shown at the top of the next page. Click **OK**.

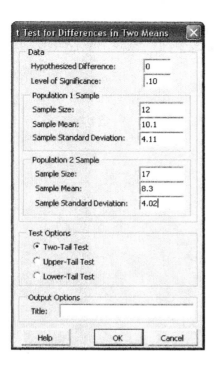

The output is displayed in a new worksheet named "Hypothesis." You may need to scroll down a couple rows to see all of it.

	A	B
1	t Test for Differences in Two Means	
2		
3	Data	
4	Hypothesized Difference	0
5	Level of Significance	0.1
6	Population 1 Sample	
7	Sample Size	12
8	Sample Mean	10.1
9	Sample Standard Deviation	4.11
10	Population 2 Sample	
11	Sample Size	17
12	Sample Mean	8.3
13	Sample Standard Deviation	4.02
14		
15	Intermediate Calculations	
16	Population 1 Sample Degrees of Freedom	11
17	Population 2 Sample Degrees of Freedom	16
18	Total Degrees of Freedom	27
19	Pooled Variance	16.4585
20	Difference in Sample Means	1.8
21	t Test Statistic	1.176775
22		
23	Two-Tail Test	
24	Lower Critical Value	-1.70329
25	Upper Critical Value	1.703288
26	p-Value	0.249556
27	Do not reject the null hypothesis	

◀

Section 8.3 Testing the Difference Between Means (Dependent Samples)

▶ Example 2 (pg. 464) | The t-Test for the Difference Between Means

1. Open a new Excel worksheet and enter the performance ratings as shown at the top of the next page.

	A	B
1	Last Year	This Year
2	60	56
3	54	48
4	78	70
5	84	60
6	91	85
7	25	40
8	50	40
9	65	55
10	68	80
11	81	75
12	75	78
13	45	50
14	62	50
15	79	85
16	58	53
17	63	60

2. At the top of the screen, select **Data** and **Data Analysis**. Select **t-Test: Paired Two Sample for Means**. Click **OK**.

If Data Analysis does not appear as a choice in the Data ribbon, you will need to load the Microsoft Excel ToolPak add-in. Follow the procedure in Section GS 8.1 before continuing.

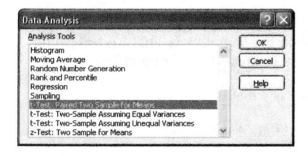

3. Complete the t-Test: Paired Two Sample for Means dialog box as shown below. Be sure to select **Labels** and change **Alpha** to .01. Click **OK**.

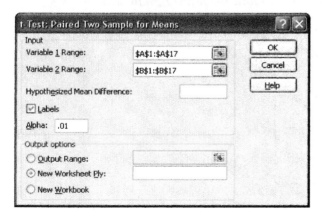

The output is displayed in a new worksheet. You will want to adjust the width of the columns so that you can read all the labels. Your output should look similar to the output displayed below.

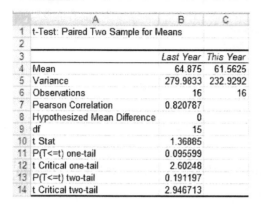

	A	B	C
1	t-Test: Paired Two Sample for Means		
2			
3		Last Year	This Year
4	Mean	64.875	61.5625
5	Variance	279.9833	232.9292
6	Observations	16	16
7	Pearson Correlation	0.820787	
8	Hypothesized Mean Difference	0	
9	df	15	
10	t Stat	1.36885	
11	P(T<=t) one-tail	0.095599	
12	t Critical one-tail	2.60248	
13	P(T<=t) two-tail	0.191197	
14	t Critical two-tail	2.946713	

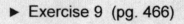

► Exercise 9 (pg. 466)	Testing the Hypothesis That Verbal SAT Scores Improved the Second Time

1. Open worksheet "Ex8_3-9" in the Chapter 8 folder.

2. At the top of the screen, select **Data** and **Data Analysis**. Select **t-Test: Paired Two Sample for Means**. Click **OK**.

If Data Analysis does not appear as a choice in the Data ribbon, you will need to load the Microsoft Excel ToolPak add-in. Follow the procedure in Section GS 8.1 before continuing.

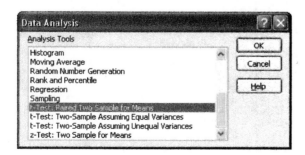

3. Complete the t-Test: Paired Two Sample for Means dialog box as shown below. Click **OK**.

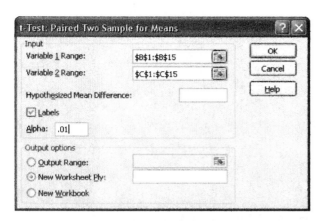

The output is displayed in a new worksheet. You will want to make the columns wider so that you can read all the labels. Your output should appear similar to the output shown below.

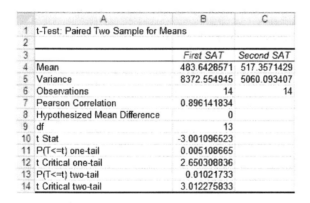

◀

| ▶ Exercise 21 (pg. 470) | Constructing a 90% Confidence Interval for μ_D, the Mean Increase in Hours of Sleep |

1. Open worksheet "Ex8_3-21" in the Chapter 8 folder.

2. Use Excel to calculate a difference score for each of the 16 patients. First, click in cell **C2** and key in the label **Difference**.

	A	B	C
1		Hours of sleep	
2	Without drug	With new drug	Difference
3	1.8	3.0	
4	2.0	3.6	
5	3.4	4.0	
6	3.5	4.4	
7	3.7	4.5	
8	3.8	5.2	
9	3.9	5.5	
10	3.9	5.7	
11	4.0	6.2	
12	4.9	6.3	
13	5.1	6.6	
14	5.2	7.8	
15	5.0	7.2	
16	4.5	6.5	
17	4.2	5.6	
18	4.7	5.9	

3. Click in cell C3 and enter the formula **=A3-B3** as shown below. Press [**Enter**].

	A	B	C
1		Hours of sleep	
2	Without drug	With new drug	Difference
3	1.8	3.0	=A3-B3

4. Click in cell **C3** (where –1.2 now appears) and copy the contents of that cell to cells C4 through C18.

	A	B	C
1		Hours of sleep	
2	Without drug	With new drug	Difference
3	1.8	3.0	-1.2
4	2.0	3.6	-1.6
5	3.4	4.0	-0.6
6	3.5	4.4	-0.9
7	3.7	4.5	-0.8
8	3.8	5.2	-1.4
9	3.9	5.5	-1.6
10	3.9	5.7	-1.8
11	4.0	6.2	-2.2
12	4.9	6.3	-1.4
13	5.1	6.6	-1.5
14	5.2	7.8	-2.6
15	5.0	7.2	-2.2
16	4.5	6.5	-2.0
17	4.2	5.6	-1.4
18	4.7	5.9	-1.2

5. You will now obtain descriptive statistics for Difference. At the top of the screen, select **Data** and **Data Analysis**. Select **Descriptive Statistics**. Click **OK**.

If Data Analysis does not appear as a choice in the Data ribbon, you will need to load the Microsoft Excel ToolPak add-in. Follow the procedure in Section GS 8.1 before continuing.

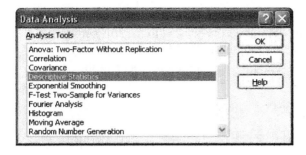

6. Complete the Descriptive Statistics dialog box as shown below. Click **OK**.

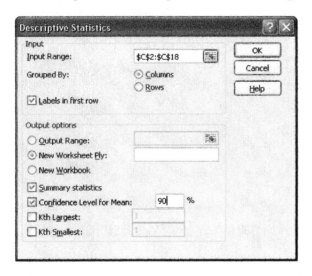

The output is displayed in a new worksheet. The mean is printed at the top and the confidence level is printed at the bottom. To calculate the lower limit of the 90% confidence interval, subtract 0.2376 from -1.525. To calculate the upper limit, add 0.2376 to -1.525.

	A	B
1	Difference	
2		
3	Mean	-1.525
4	Standard Error	0.135554
5	Median	-1.45
6	Mode	-1.6
7	Standard Deviation	0.542218
8	Sample Variance	0.294
9	Kurtosis	-0.26069
10	Skewness	-0.23658
11	Range	2
12	Minimum	-2.6
13	Maximum	-0.6
14	Sum	-24.4
15	Count	16
16	Confidence Level(90.0%)	0.237634

◀

Section 8.4 Testing the Difference Between Proportions

▶ Example 1 (pg. 473)	A Two-Sample z-Test for the Difference Between Proportions

If the PHStat add-in has not been loaded, you will need to load it before continuing. Follow the instructions in Section GS 8.2.

1. First open a new Excel worksheet. Then, at the top of the screen, select **Add-Ins** and **PHStat**. Select **Two-Sample Tests→Z Test for Differences in Two Proportions**.

2. Complete the Z Test for Differences in Two Proportions dialog box as shown below. Click **OK**.

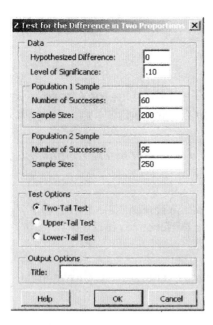

The output is displayed in a worksheet named "Hypothesis."

	A	B
1	**Z Test for Differences in Two Proportions**	
2		
3	**Data**	
4	**Hypothesized Difference**	0
5	**Level of Significance**	0.1
6	**Group 1**	
7	**Number of Successes**	60
8	**Sample Size**	200
9	**Group 2**	
10	**Number of Successes**	95
11	**Sample Size**	250
12		
13	**Intermediate Calculations**	
14	Group 1 Proportion	0.3
15	Group 2 Proportion	0.38
16	Difference in Two Proportions	-0.08
17	Average Proportion	0.344444444
18	Z Test Statistic	-1.774615984
19		
20	**Two-Tail Test**	
21	**Lower Critical Value**	-1.644853627
22	**Upper Critical Value**	1.644853627
23	***p*-Value**	0.075961316
24	**Reject the null hypothesis**	

| ► Exercise 7 (pg. 475) | Testing the Claim That the Use of Alternative Medicines Has Not Changed |

If the PHStat add-in has not been loaded, you will need to load it before continuing. Follow the instructions in Section GS 8.2.

---

1. First open a new Excel worksheet. Then, at the top of the screen, select **Add-Ins** and **PHStat**. Select **Two-Sample Tests→Z Test for Differences in Two Proportions**.

2. Complete the Z Test for Differences in Two Proportions dialog box as shown at the top of the next page. Click **OK**.

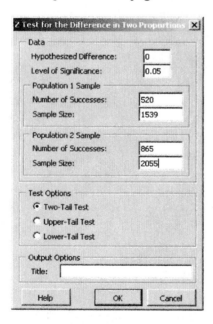

The output is displayed in a worksheet named "Hypothesis."

	A	B
1	Z Test for Differences in Two Proportions	
2		
3	Data	
4	Hypothesized Difference	0
5	Level of Significance	0.05
6	Group 1	
7	Number of Successes	520
8	Sample Size	1539
9	Group 2	
10	Number of Successes	865
11	Sample Size	2055
12		
13	Intermediate Calculations	
14	Group 1 Proportion	0.337881741
15	Group 2 Proportion	0.420924574
16	Difference in Two Proportions	-0.083042833
17	Average Proportion	0.385364496
18	Z Test Statistic	-5.06166817
19		
20	Two-Tail Test	
21	Lower Critical Value	-1.959963985
22	Upper Critical Value	1.959963985
23	p-Value	4.15604E-07
24	Reject the null hypothesis	

◀

Technology

▶ Exercise 1 (pg. 487)	Testing the Hypothesis That the Probability of a "Found Coin" Lying Heads Up Is 0.5

If the PHStat add-in has not been loaded, you will need to load it before continuing. Follow the instructions in Section GS 8.2.

1. First open a new Excel worksheet. Then, at the top of the screen, select **Add-Ins** and **PHStat**. Select **One-Sample Tests→Z Test for the Proportion**.

2. Complete the Z Test for the Proportion dialog box as shown at the top of the next page. Click **OK**.

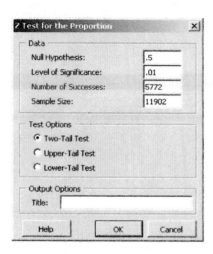

Your output should look similar to the output displayed below.

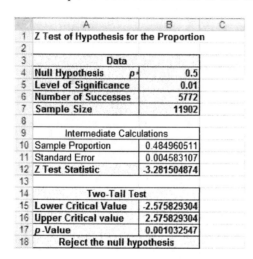

	A	B	C
1	Z Test of Hypothesis for the Proportion		
2			
3	Data		
4	Null Hypothesis p =	0.5	
5	Level of Significance	0.01	
6	Number of Successes	5772	
7	Sample Size	11902	
8			
9	Intermediate Calculations		
10	Sample Proportion	0.484960511	
11	Standard Error	0.004583107	
12	Z Test Statistic	-3.281504874	
13			
14	Two-Tail Test		
15	Lower Critical Value	-2.575829304	
16	Upper Critical value	2.575829304	
17	p-Value	0.001032547	
18	Reject the null hypothesis		

► Exercise 3 (pg. 487)   Simulating "Tails Over Heads"

1. Open a new Excel worksheet.

2. At the top of the screen, select **Data** and **Data Analysis**. Select **Random Number Generation**. Click **OK**.

If Data Analysis does not appear as a choice in the Data ribbon, you will need to load the Microsoft Excel ToolPak add-in. Follow the procedure in Section GS 8.1 before continuing.

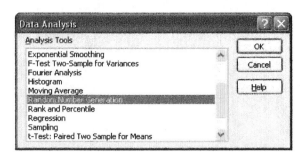

Complete the Random Number Generation dialog box as shown below. The output will be placed in cell A1. Click **OK**.

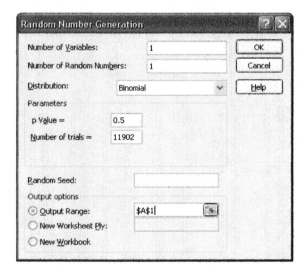

The output for this simulation indicates that 5,930 of the 11,902 coins were found heads up. Because this output was generated randomly, it is unlikely that your output will be exactly the same.

3. Complete the Descriptive Statistics dialog box as shown below. Be sure to select **Labels in First Row** and **Summary statistics**. Click **OK**.

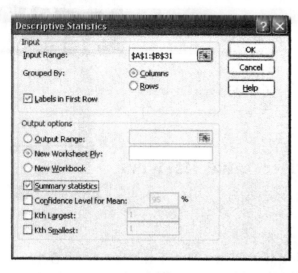

4. You will want to make column A and column C wider so that the labels are easier to read. The variance of Time A is 0.2401 and the variance of Time B is 0.2212. Return to the worksheet that contains the data. To do this, click on the **Sheet1** tab near the bottom of the screen.

	A	B	C	D
1	Time A		Time B	
2				
3	Mean	2.13	Mean	1.756667
4	Standard Error	0.089462	Standard Error	0.085861
5	Median	2.1	Median	1.7
6	Mode	2.1	Mode	1.6
7	Standard Deviation	0.490004	Standard Deviation	0.470277
8	Sample Variance	0.240103	Sample Variance	0.221161
9	Kurtosis	0.863836	Kurtosis	0.200666
10	Skewness	-0.00957	Skewness	0.531138
11	Range	2.3	Range	2
12	Minimum	1	Minimum	0.9
13	Maximum	3.3	Maximum	2.9
14	Sum	63.9	Sum	52.7
15	Count	30	Count	30

Correlation and Regression

Section 9.1 Correlation

| ▶ Example 3 (pg. 498) | Constructing a Scatter Plot Using Technology |

1. Open worksheet "Old Faithful" in the Chapter 9 folder.

2. Click on any cell within the table of data. At the top of the screen, click **Insert** and select **Scatter** from the Charts group.

3. In the scatter type box, select the leftmost chart in the top row by clicking on it.

4. In the Design ribbon, click the leftmost diagram (**Layout 1**) in Chart Layouts so that you can add an X-axis title, a Y-axis title, and revise the main chart title.

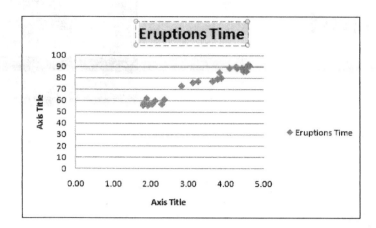

5. Click on **Eruptions Time** and replace it with **Duration of Old Faithful's Eruptions by Time Until the Next Eruption**. Click on the Y **Axis Title** and replace it with **Time Until the Next Eruption (in minutes)**. Click on the X **Axis Title** and replace it with **Duration (in minutes)**. `

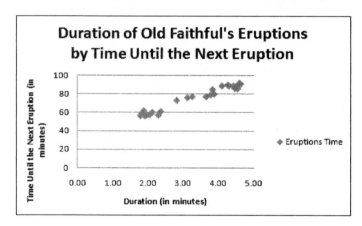

6. To remove the legend on the right, click on **Eruptions Time** and press [**Delete**].

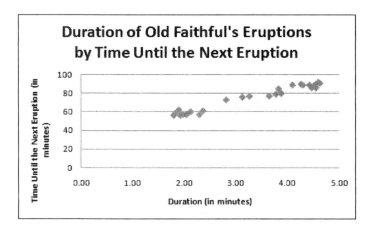

7. Let's move the scatter plot to a new worksheet. To do this, **right-click** in the white area near a border and select **Move Chart** from the menu that appears.

8. Select **New sheet** and click **OK**.

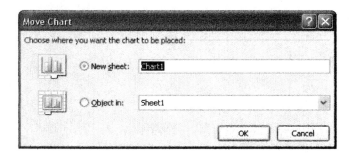

Your scatter plot should look similar to the one shown below.

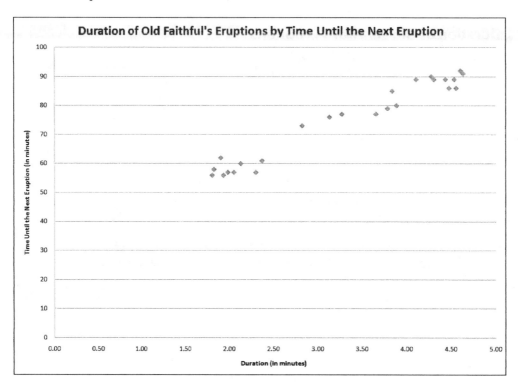

Duration of Old Faithful's Eruptions by Time Until the Next Eruption

◀

▶ Example 5 (pg. 501)

Using Technology to Find a Correlation Coefficient

1. Open worksheet "Old Faithful" in the Chapter 9 folder and click in cell **C1** to place the output there.

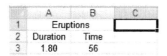

	A	B	C
1	Eruptions		
2	Duration	Time	
3	1.80	56	

2. At the top of the screen, select **Formulas** and **Insert Function**.

3. Select the **Statistical** category. Select the **CORREL** function. Click **OK**.

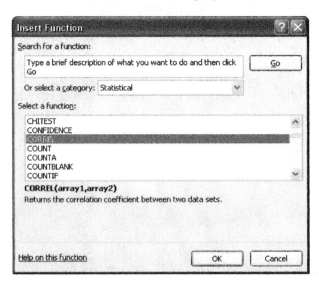

4. Complete the CORREL dialog box as shown below. Click **OK**.

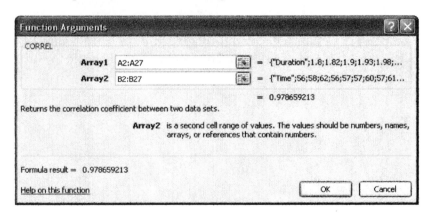

Your output should look like the output shown below.

	A	B	C
1	Eruptions		0.978659
2	Duration	Time	
3	1.80	56	

▶ Exercise 17 (pg. 508) | **Scatter Plot and Correlation for Hours Studying and Test Scores**

1. Open worksheet "Ex9_1-17" in the Chapter 9 folder.

2. Click on any cell within the table of data. At the top of the screen, click **Insert** and select **Scatter** in the Charts group.

3. In the scatter type box, select the leftmost chart in the top row by clicking on it.

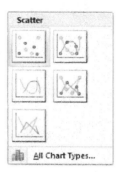

4. In the Design ribbon, click the leftmost diagram (**Layout 1**) from Chart Layouts so that you can add an X-axis title, a Y-axis title, and revise the main chart title.

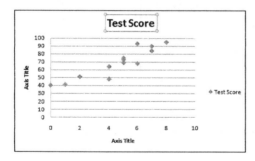

5. Click on **Test Score** and replace it with **Hours Spent Studying for a Test by Test Score**. Click on the Y **Axis Title** and replace it with **Test Score**. Click on the X **Axis Title** and replace it with **Hours**.

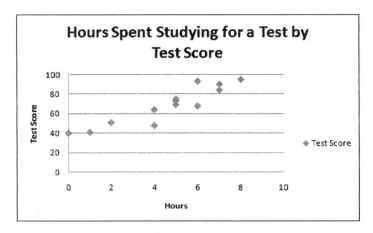

6. To remove the legend on the right, click on **Test Score** and press [**Delete**].

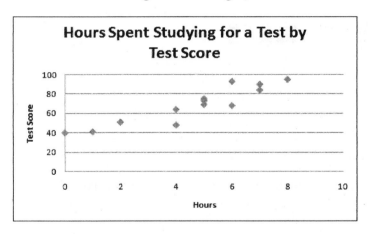

7. Let's move the scatter plot to a new worksheet. To do this, **right click** in the white area near a border and select **Move Chart** from the menu that appears.

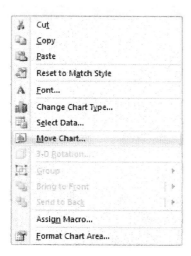

8. Select **New sheet** and click **OK**.

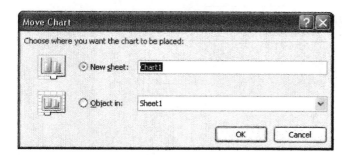

Your scatter plot should look similar to the one shown at the top of the next page.

9. You will now use Excel to find the correlation between hours and test score. Return to the worksheet containing the data by clicking the **Sheet1** tab near the bottom of the screen. Click in cell **C1** of the worksheet to place the output there.

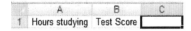

10. At the top of the screen, select **Formulas** and **Insert Function**.

11.   Select the **Statistical** category. Select the **CORREL** function. Click **OK**.

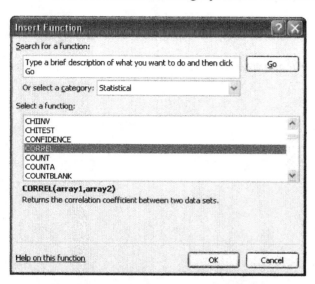

12.   Complete the CORREL dialog box as shown below. Click **OK**.

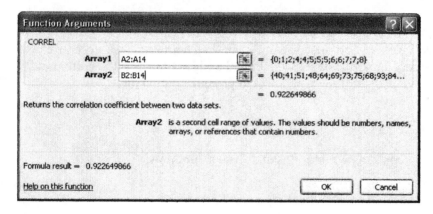

Your output should look like the output shown below.

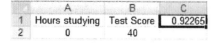

Section 9.2 Linear Regression

▶ Example 2 (pg. 515) | Using Technology to Find a Regression Equation

1. Open worksheet "Old Faithful" in the Chapter 9 folder.

2. You will be using Excel's functions to obtain the slope and intercept. Type the labels **Slope** and **Intercept** in the worksheet as shown below. Then click in cell **D1** where the slope will be placed.

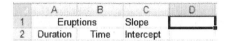

3. At the top of the screen, select **Formulas** and **Insert Function**.

4. Select the **Statistical** category. Select the **SLOPE** function. Click **OK**.

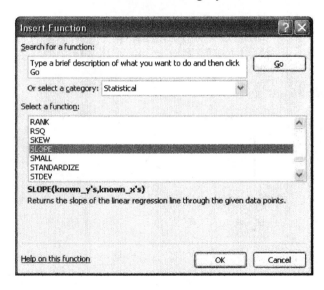

5. Complete the SLOPE dialog box as shown below. Click **OK**.

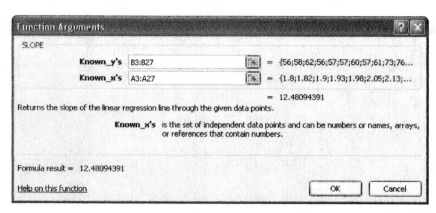

6. A slope of 12.48 is returned and placed in cell D1 of the worksheet. Click in cell **D2** where the intercept will be placed.

7. At the top of the screen select **Formulas** and **Insert Function**.

8. Select the **Statistical** category. Select the **INTERCEPT** function. Click **OK**.

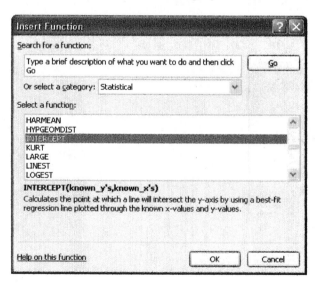

9. Complete the INTERCEPT dialog box as shown below. Click **OK**.

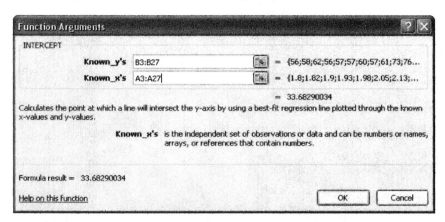

An intercept of 33.68 is returned and placed in cell D2 of the worksheet.

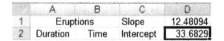

◀

| | Exercise 16 (pg. 518) | Finding the Equation of a Regression Line for Hours Online and Test Scores |

1. Open worksheet "Ex9_2-16" in the Chapter 9 folder.

2. At the top of the screen, select **Data** and **Data Analysis**. Select **Regression** and click **OK**.

If Data Analysis does not appear as a choice in the Data ribbon, you will need to load the Microsoft Excel Analysis ToolPak add-in. Follow the procedure in Section GS 8.1 before continuing.

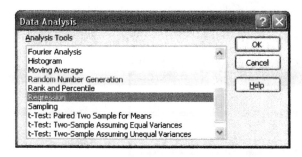

3.   Complete the Regression dialog box as shown below. Click **OK**.

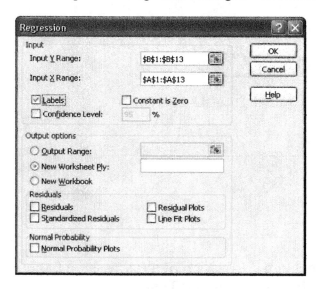

The intercept and slope of the regression equation are shown in the bottom two lines of the output under the label "Coefficients." The intercept is 93.97 and the slope is – 4.07.

16		Coefficients	Standard Error	t Stat	P-value	Lower 95%	Upper 95%
17	Intercept	93.97003745	4.523586141	20.77335	1.48E-09	83.890859	104.049215
18	Hours onli	-4.06741573	0.860011871	-4.72949	0.000805	-5.983642	-2.1511899

Section 9.3 Measures of Regression and Prediction Intervals

▶ Example 2 (pg. 528)	Finding the Standard Error of Estimate

1. Open a new Excel worksheet and enter the expenses and sales data as shown below.

	A	B
1	Expenses	Sales
2	2.4	225
3	1.6	184
4	2	220
5	2.6	240
6	1.4	180
7	1.6	184
8	2	186
9	2.2	215

2. At the top of the screen, select **Data** and **Data Analysis**. Select **Regression** and click **OK**.

If Data Analysis does not appear as a choice in the Data ribbon, you will need to load the Microsoft Excel Analysis ToolPak add-in. Follow the procedure in Section GS 8.1 before continuing.

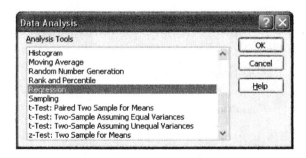

3. Complete the Regression dialog box as shown below. If you select **Residuals**, the output will include $y - \hat{y}$ for each observation in the data set. Click **OK**.

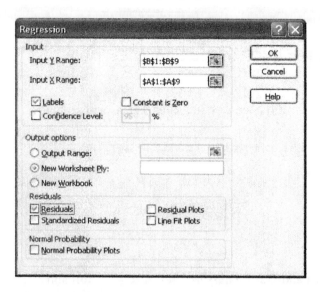

The standard error of estimate is displayed in the top part of the output. You will want to widen column A so that you can read all the labels. The standard error of estimate is equal to 10.29.

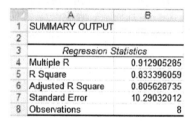

	A	B
1	SUMMARY OUTPUT	
2		
3	*Regression Statistics*	
4	Multiple R	0.912905285
5	R Square	0.833396059
6	Adjusted R Square	0.805628735
7	Standard Error	10.29032012
8	Observations	8

The residuals are displayed in the lower part of the output.

	Observation	Predicted Sales	Residuals
22	RESIDUAL OUTPUT		
23			
24	*Observation*	*Predicted Sales*	*Residuals*
25	1	225.8097166	-0.809716599
26	2	185.2267206	-1.226720648
27	3	205.5182186	14.48178138
28	4	235.9554656	4.044534413
29	5	175.0809717	4.91902834
30	6	185.2267206	-1.226720648
31	7	205.5182186	-19.51821862
32	8	215.6639676	-0.663967611

Section 9.4 Multiple Regression

▶ Example 1 (pg. 536) | Finding a Multiple Regression Equation

1. Open worksheet "Salary" in the Chapter 9 folder.

2. At the top of the screen, select **Data** and **Data Analysis**. Select **Regression** and click **OK**.

If Data Analysis does not appear as a choice in the Data ribbon, you will need to load the Microsoft Excel Analysis ToolPak add-in. Follow the procedure in Section GS 8.1 before continuing.

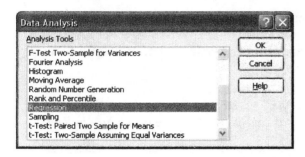

3. Complete the Regression dialog box as shown below. Click **OK**.

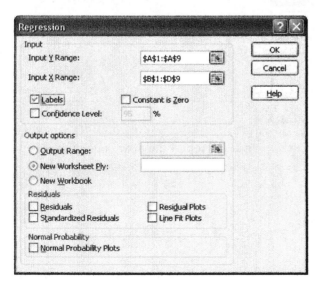

The coefficients for the multiple regression equation are displayed in the lower portion of the output under the label "Coefficients."

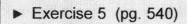

16		Coefficients	Standard Error	t Stat	P-value	Lower 95%
17	Intercept	45764.448	1581.346465	28.11848	1.49E-05	44283.35
18	Employment (yrs)	364.4120281	48.31750816	7.542029	0.001656	230.2611
19	Experience (yrs)	227.6188106	123.8361513	1.838064	0.139912	-116.2055
20	Education (yrs)	266.9350412	147.3556227	1.811502	0.144295	-142.1898

◀

▶ Exercise 5 (pg. 540)	Finding a Multiple Regression Equation, the Standard Error of Estimate, and R^2

1.   Open worksheet "Ex9_4-5" in the Chapter 9 folder.

2.   At the top of the screen, select **Data** and **Data Analysis**. Selection **Regression** and click **OK**.

If Data Analysis does not appear as a choice in the Data ribbon, you will need to load the Microsoft Excel Analysis ToolPak add-in. Follow the procedure in Section GS 8.1 before continuing.

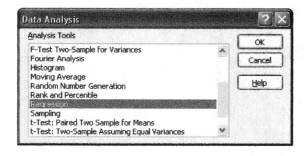

3.   Complete the Regression dialog box as shown at the top of the next page. Click **OK**.

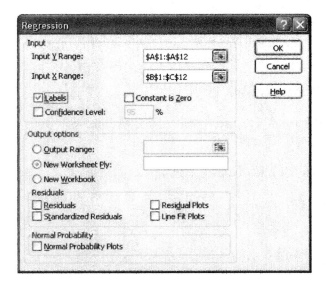

R^2 is equal to 0.9850. The standard error of estimate is equal to 28.489. The coefficients for the multiple regression equation are displayed in the lower portion of the output under the label "Coefficients."

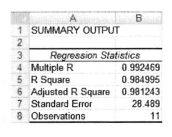

	A	B
1	SUMMARY OUTPUT	
2		
3	Regression Statistics	
4	Multiple R	0.992469
5	R Square	0.984995
6	Adjusted R Square	0.981243
7	Standard Error	28.489
8	Observations	11

16		Coefficients
17	Intercept	-2518.36355
18	Square footage	126.8217972
19	Shopping centers	66.35999407

◄

Technology

▶ Exercise 1 (pg. 549)   Constructing a Scatter Plot

If the PHStat add-in has not been loaded, you will need to load it before continuing. Follow the instructions in Section GS 8.2.

Directions are given here for constructing a scatter plot of the calories and sugar variables. You can follow these instructions for constructing scatter plots of the other pairs of variables, but you will need to change the data ranges.

1.   Open worksheet "Tech9" in the Chapter 9 folder.

2.   At the top of the screen, select **Add-Ins** and **PHStat**. Select **Regression→Simple Linear Regression**.

3.   Complete the Simple Linear Regression dialog box as shown below. Click **OK**.

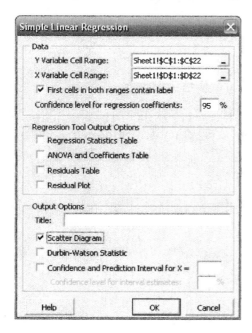

4.   To see the scatter plot, click the **Scatter** sheet tab near the bottom of the screen. There is a great deal of blank space in the lower half of the scatter plot. You will adjust this by changing the Y-axis scale. **Right-click** on any number on the Y-axis scale. I clicked on 60.

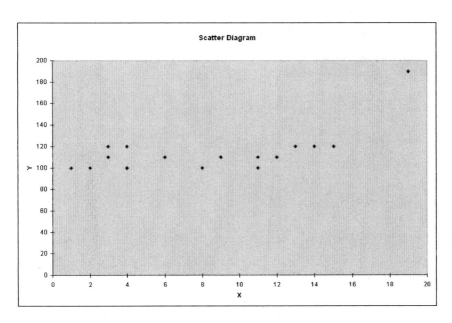

5. Select **Format Axis** from the shortcut menu that appears.

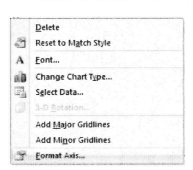

6.   Change Minimum to Fixed **80**.   Click **Close**.

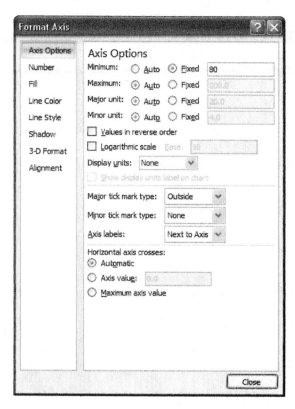

The completed scatter plot is shown at the top of the next page.

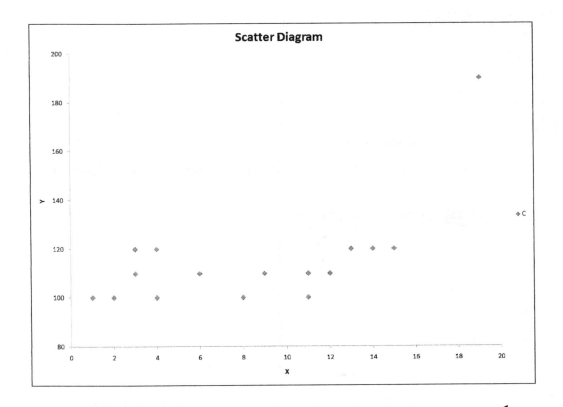

► Exercise 3 (pg. 549) Find the Correlation Coefficient for Each Pair of Variables

1.   Open worksheet "Tech9" in the Chapter 9 folder.

*If you have just completed Exercise 1 on page 549 and have not closed the Excel worksheet, return to the sheet containing the data by clicking on the **Sheet1** tab at the bottom of the screen.*

2.   At the top of the screen, select **Data** and **Data Analysis**. Select **Correlation** and click **OK**.

If Data Analysis does not appear as a choice in the Data ribbon, you will need to load the Microsoft Excel Analysis ToolPak add-in. Follow the procedure in Section GS 8.1 before continuing.

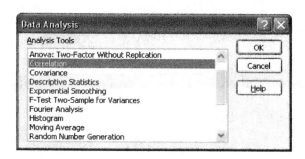

3. Complete the Correlation dialog box as shown below. Be sure to select **Labels in First Row**.   Click **OK**.

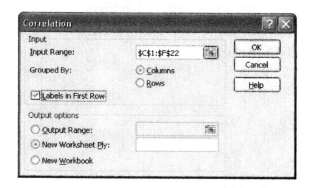

The output is a correlation matrix that displays the correlation coefficients for all pairs of variables.

	A	B	C	D	E
1		C	S	F	R
2	C		1			
3	S	0.565937	1		
4	F	0.194262	0.311011	1	
5	R	0.926983	0.543735	-0.02056	1

▶ Exercise 4 (pg. 549)   **Finding the Equation of a Regression Line**

Directions are given here for finding the equation for calories (X) and sugar (Y).

1. Complete Exercise 1 on page 549 following the instructions given in this manual. Your output is a scatter plot of the calories and sugar variables.

2. **Right-click** on one of the dots and select **Add Trendline**.

3. Select the **Linear** option. At the bottom of the dialog box, select **Display Equation on chart** and **Display R-squared value on chart**. Click **Close**.

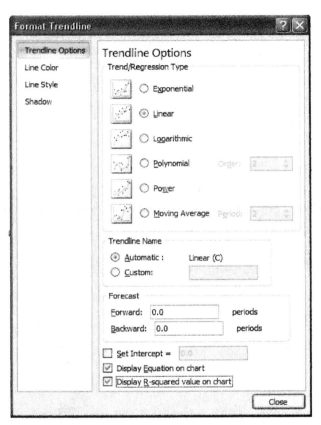

4. Move the equation to a location on the scatter plot where it can be easily read. The slope is 2.115, and the intercept is 96.35.

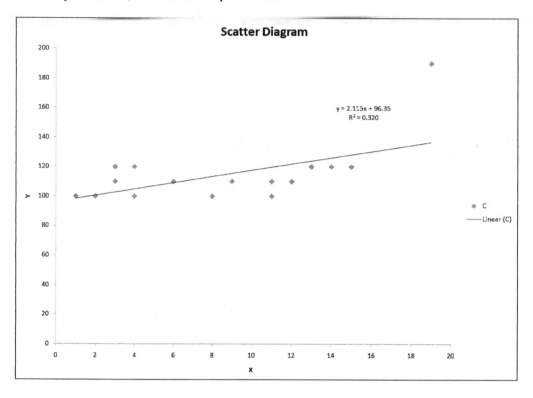

► Exercise 6 (pg. 549)   Finding Multiple Regression Equations

Directions are given here for Exercise 6 (a) $C = b + m_1S + m_2F + m_3R$.

1. Open worksheet "Tech9" in the Chapter 9 folder.

---

*If you have just completed Exercise 1, Exercise 3, or Exercise 4 on page 549 and have not closed the Excel worksheet, return to the sheet containing the data by clicking on the **Sheet1** tab at the bottom of the screen.*

---

2. At the top of the screen, select **Data** and **Data Analysis**. Select **Regression** and click **OK**.

If Data Analysis does not appear as a choice in the Data ribbon, you will need to load the Microsoft Excel Analysis ToolPak add-in. Follow the procedure in Section GS 8.1 before continuing.

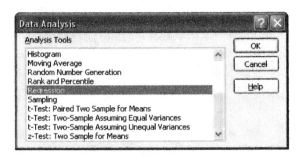

3.   Complete the Regression dialog box as shown below. Click **OK**.

The predictor variables must be located in adjacent columns in the Excel worksheet.

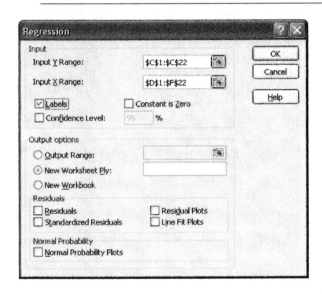

The coefficients for the equation are given in the lower part of the output.

16		Coefficients
17	Intercept	22.05502364
18	S	-0.042670433
19	F	6.643653591
20	R	3.455152976

Chi-Square Tests and the F-Distribution

Section 10.2 Independence

▶ Example 3 (pg. 570)	Using Technology for a Chi-Square Independence Test

If the PHStat add-in has not been loaded, you will need to load it before continuing. Follow the instructions in Section GS 8.2.

1. First, open a new Excel worksheet. Then, at the top of the screen, select **Add-Ins** and **PHStat**. Then select **Multiple-Sample Tests→Chi-Square Test**.

2. Complete the Chi-Square Test dialog box as shown below. Click **OK**.

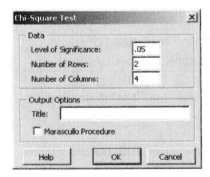

3. A Chi-Square Test worksheet will appear. Delete the user note, and then complete the Observed Frequencies table in this worksheet as shown at the top of the next page. You will notice that values in other locations of the worksheet change as you enter the observed frequencies.

	A	B	C	D	E	F
1	Chi-Square Test					
2						
3		Observed Frequencies				
4			Column variable			
5	Row variable	C1	C2	C3	C4	Total
6	R1	40	53	26	6	125
7	R2	34	68	37	11	150
8	Total	74	121	63	17	275

The critical value, the obtained chi-square test statistic, and the p-value are displayed in rows 24 through 26. The statistical decision is displayed in row 27.

17	Data
18	Level of Significance 0.05
19	Number of Rows 2
20	Number of Columns 4
21	Degrees of Freedom 3
22	
23	Results
24	Critical Value 7.814728
25	Chi-Square Test Statistic 3.493357
26	p-Value 0.321625
27	Do not reject the null hypothesis
28	
29	Expected frequency assumption
30	is met.

◀

▶ Exercise 17 (pg. 573)	Testing the Drug for Treatment of Obsessive-Compulsive Disorder

If the PHStat add-in has not been loaded, you will need to load it before continuing. Follow the instructions in Section GS 8.2.

1. Open a new Excel worksheet. At the top of the screen, select **Add-Ins** and **PHStat**. Select **Multiple-Sample Tests→Chi-Square Test**.

2. Complete the Chi-Square Test dialog box as shown at the top of the next page. Click **OK**.

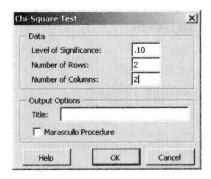

3. A Chi-Square Test worksheet will appear. Complete the Observed Frequencies
    table in this worksheet as shown below. Values in other locations of the worksheet
    change as you enter the observed frequencies.

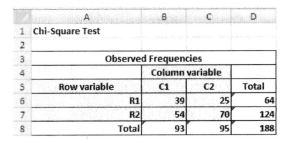

	A	B	C	D
1	Chi-Square Test			
2				
3		Observed Frequencies		
4		Column variable		
5	Row variable	C1	C2	Total
6	R1	39	25	64
7	R2	54	70	124
8	Total	93	95	188

The critical value, the obtained chi-square test statistic, and the p-value are displayed in
rows 24 through 26. The statistical decision is displayed in row 27.

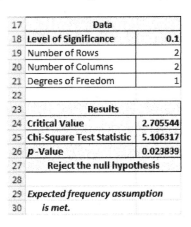

17	Data	
18	Level of Significance	0.1
19	Number of Rows	2
20	Number of Columns	2
21	Degrees of Freedom	1
22		
23	Results	
24	Critical Value	2.705544
25	Chi-Square Test Statistic	5.106317
26	p-Value	0.023839
27	Reject the null hypothesis	
28		
29	Expected frequency assumption	
30	is met.	

Section 10.3 Comparing Two Variances

▶ Example 3 (pg. 583)

Performing a Two-Sample F-Test

The example problem asks you to compare the variances of the time customers wait before their meals are served.

If the PHStat add-in has not been loaded, you will need to load it before continuing. Follow the instructions in Section GS 8.2.

---

1. First open a new Excel worksheet. Then, at the top of the screen, select **Add-Ins** and **PHStat**. Select **Two-Sample Tests→F Test for Differences in Two Variances**.

2. Complete the F Test for Differences in Two Variances dialog box as shown below. Click **OK**.

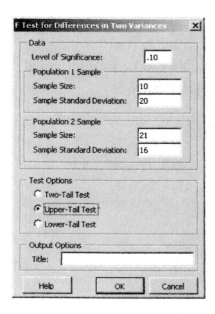

Your output should appear similar to the output displayed at the top of the next page.

	A	B	C	D	E
1	F Test for Differences in Two Variances				
2					
3	Data				
4	Level of Significance	0.1			
5	Population 1 Sample				
6	Sample Size	10			
7	Sample Standard Deviation	20			
8	Population 2 Sample				
9	Sample Size	21			
10	Sample Standard Deviation	16			
11					
12	Intermediate Calculations				
13	F Test Statistic	1.5625			
14	Population 1 Sample Degrees of Freedom	9			
15	Population 2 Sample Degrees of Freedom	20			
16				Calculations Area	
17	Upper-Tail Test			FDIST value	0.193904
18	Upper Critical Value	1.964853		1-FDIST value	0.806096
19	p-Value	0.193904			
20	Do not reject the null hypothesis				

Section 10.4 Analysis of Variance

▶ Example 2 (pg. 593)	Using Technology to Perform ANOVA Tests

1. Open worksheet "Airlines" in the Chapter 10 folder.

2. At the top of the screen, select **Data** and **Data Analysis**. Select **ANOVA: Single Factor** and click **OK**.

If Data Analysis does not appear as a choice in the Data ribbon, you will need to load the Microsoft Excel Analysis ToolPak add-in. Follow the procedure in Section GS 8.1 before continuing.

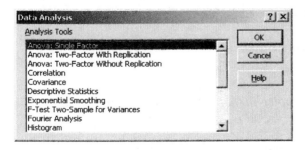

3. Complete the ANOVA: Single Factor dialog box as shown at the top of the next page. Be sure to select **Labels in first row** and change **Alpha** to **.01**. Click **OK**.

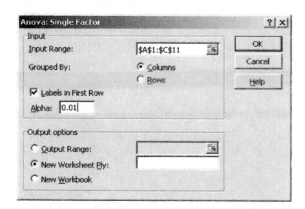

Make column A wider so that you can read all the labels. Your output should appear similar to the output displayed below.

	A	B	C	D	E	F	G
1	Anova: Single Factor						
2							
3	SUMMARY						
4	Groups	Count	Sum	Average	Variance		
5	Airline 1	10	1238	123.8	106.6222		
6	Airline 2	10	1315	131.5	120.2778		
7	Airline 3	10	1427	142.7	215.1222		
8							
9							
10	ANOVA						
11	Source of Variation	SS	df	MS	F	P-value	F crit
12	Between Groups	1806.467	2	903.2333	6.130235	0.006383	5.488118
13	Within Groups	3978.2	27	147.3407			
14							
15	Total	5784.667	29				

◀

▶ **Exercise 5 (pg. 595)**   **Testing the Claim That the Mean Toothpaste Costs Per Ounce Are Different**

1. Open worksheet "Ex10_4-5" in the Chapter 10 folder.

2. At the top of the screen, select **Data** and **Data Analysis**. Select **ANOVA: Single Factor** and click **OK**.

If Data Analysis does not appear as a choice in the Data ribbon, you will need to load the Microsoft Excel Analysis ToolPak add-in. Follow the procedure in Section GS 8.1 before continuing.

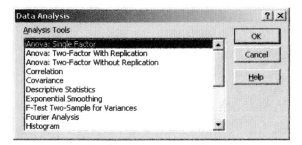

3. Complete the ANOVA: Single Factor dialog box as shown below. Click **OK**.

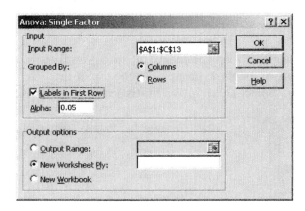

Make column A wider so that you can read all the labels. Your output should appear similar to the output displayed below.

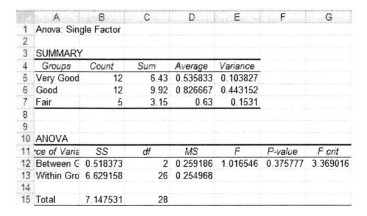

	A	B	C	D	E	F	G
1	Anova: Single Factor						
2							
3	SUMMARY						
4	Groups	Count	Sum	Average	Variance		
5	Very Good	12	6.43	0.535833	0.103827		
6	Good	12	9.92	0.826667	0.443152		
7	Fair	5	3.15	0.63	0.1531		
8							
9							
10	ANOVA						
11	rce of Varia	SS	df	MS	F	P-value	F crit
12	Between G	0.518373	2	0.259186	1.016546	0.375777	3.369016
13	Within Gro	6.629158	26	0.254968			
14							
15	Total	7.147531	28				

Technology

► Exercise 3 (pg. 609)	Determining Whether the Populations Have Equal Variances—Teacher Salaries

If the PHStat add-in has not been loaded, you will need to load it before continuing. Follow the instructions in Section GS 8.2.

1. Open worksheet "Tech10-a" in the Chapter 10 folder.

2. Because you are working with raw data rather than summary statistics, you first need to calculate the standard deviations of the three groups. At the top of the screen, select **Data** and **Data Analysis**. Select **Descriptive Statistics** and click **OK**.

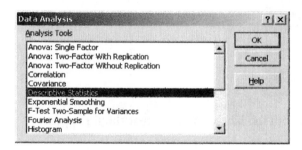

3. Complete the Descriptive Statistics dialog box as shown below. Click **OK**.

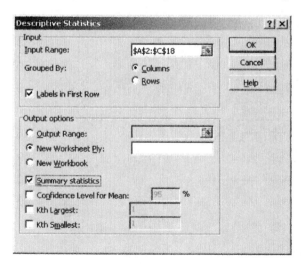

4. Make the columns wider so that you can read the labels. The standard deviations of California, Ohio, and Wyoming salaries are 7664.97, 7847.782, and 7887.628,

respectively. For this exercise, instructions are provided only for testing the equality of the California and Ohio salaries. At the top of the screen, select **Add-Ins** and **PHStat**. Select **Two-Sample Tests→F Test for Differences in Two Variances**.

	A	B	C	D	E	F
1	California		Ohio		Wyoming	
2						
3	Mean	57604.25	Mean	49438.5	Mean	40486.44
4	Standard Error	1916.243	Standard Error	1961.945	Standard Error	1971.907
5	Median	58936.5	Median	46850	Median	37157.5
6	Mode	63200	Mode	#N/A	Mode	#N/A
7	Standard Deviation	7664.97	Standard Deviation	7847.782	Standard Deviation	7887.628
8	Sample Variance	58751772	Sample Variance	61587681	Sample Variance	62214681
9	Kurtosis	-0.27133	Kurtosis	0.232469	Kurtosis	0.055315
10	Skewness	-0.54884	Skewness	0.862842	Skewness	0.980357
11	Range	27000	Range	27440	Range	25582
12	Minimum	41400	Minimum	37760	Minimum	32440
13	Maximum	68400	Maximum	65200	Maximum	58022
14	Sum	921668	Sum	791016	Sum	647783
15	Count	16	Count	16	Count	16

5. Complete the F Test for Differences in Two Variances dialog box as below. Click **OK**.

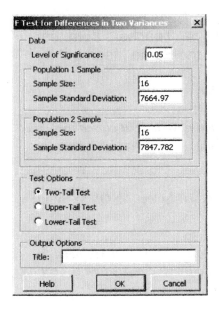

Your output should appear similar to the output displayed at the top of the next page.

	A	B	C	D	E
1	F Test for Differences in Two Variances				
2					
3	Data				
4	Level of Significance	0.05			
5	Population 1 Sample				
6	Sample Size	16			
7	Sample Standard Deviation	7664.97			
8	Population 2 Sample				
9	Sample Size	16			
10	Sample Standard Deviation	7847.782			
11					
12	Intermediate Calculations				
13	F Test Statistic	0.953953			
14	Population 1 Sample Degrees of Freedom	15			
15	Population 2 Sample Degrees of Freedom	15			
16				Calculations Area	
17	Two-Tail Test			FDIST value	0.535767
18	Lower Critical Value	0.349395		1-FDIST value	0.464233
19	Upper Critical Value	2.862093			
20	p-Value	0.928465			
21	Do not reject the null hypothesis				

◀

▶ Exercise 4 (pg. 609)	**Testing the Claim That Teachers from the Three States Have the Same Mean Salary**

1.   Open worksheet "Tech10-a" in the Chapter 10 folder.

---

*If you have just completed Exercise 3 on page 609 and have not yet closed the Excel worksheet, return to the sheet with the data by clicking on the **Sheet1** tab near the bottom of the screen.*

---

2.   At the top of the screen, select **Data** and **Data Analysis**. Select **ANOVA: Single Factor** and click **OK**.

If Data Analysis does not appear as a choice in the Data ribbon, you will need to load the Microsoft Excel Analysis ToolPak add-in. Follow the procedure in Section GS 8.1 before continuing.

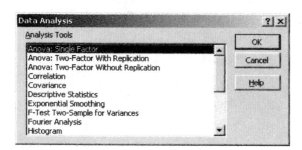

3.   Complete the ANOVA: Single Factor dialog box as shown below. Click **OK**.

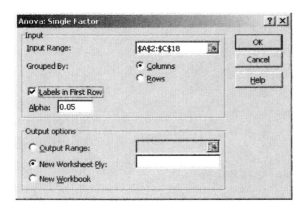

Make column A wider so that you can read all the labels. Your output should appear similar to the output shown below.

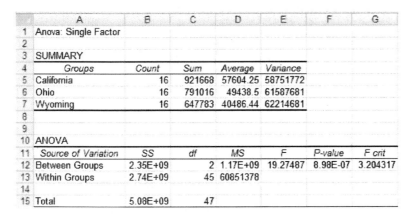

	A	B	C	D	E	F	G
1	Anova: Single Factor						
2							
3	SUMMARY						
4	*Groups*	*Count*	*Sum*	*Average*	*Variance*		
5	California	16	921668	57604.25	58751772		
6	Ohio	16	791016	49438.5	61587681		
7	Wyoming	16	647783	40486.44	62214681		
8							
9							
10	ANOVA						
11	*Source of Variation*	*SS*	*df*	*MS*	*F*	*P-value*	*F crit*
12	Between Groups	2.35E+09	2	1.17E+09	19.27487	8.98E-07	3.204317
13	Within Groups	2.74E+09	45	60851378			
14							
15	Total	5.08E+09	47				

Nonparametric Tests

Section 11.2 The Wilcoxon Tests

▶ Example 2 (pg. 627) | Performing a Wilcoxon Rank Sum Test

If the PHStat add-in has not been loaded, you will need to load it before continuing. Follow the instructions in Section GS 8.2.

1. Open worksheet "Earnings" in the Chapter 11 folder.

2. At the top of the screen, select **Add-Ins** and **PHStat**. Select **Two-Sample Tests→Wilcoxon Rank Sum Test**.

3. Complete the Wilcoxon Rank Sum Test dialog box as shown below. The Population 1 Sample Cell Range is A1 through A11. The Population 2 Sample Cell Range is B1 through B13. Click **OK**.

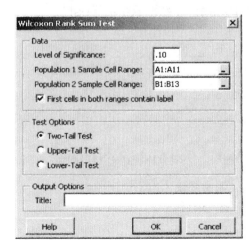

Your output should look similar to the output displayed below.

	A	B
1	Wilcoxon Rank Sum Test	
2		
3	Data	
4	Level of Significance	0.1
5		
6	Population 1 Sample	
7	Sample Size	10
8	Sum of Ranks	138
9	Population 2 Sample	
10	Sample Size	12
11	Sum of Ranks	115
12		
13	Intermediate Calculations	
14	Total Sample Size n	22
15	T1 Test Statistic	138
16	T1 Mean	115
17	Standard Error of T1	15.16575
18	Z Test Statistic	1.516575
19		
20	Two-Tail Test	
21	Lower Critical Value	-1.64485
22	Upper Critical Value	1.644854
23	p-Value	0.129374
24	Do not reject the null hypothesis	

◀

Section 11.3 The Kruskal-Wallis Test

▶ **Example 1 (pg. 635)** | Performing a Kruskal-Wallis Rank Test

If the PHStat add-in has not been loaded, you will need to load it before continuing. Follow the instructions in Section GS 8.2.

1. Open worksheet "Actuaries" in the Chapter 11 folder.

2. At the top of the screen, select **Add-Ins** and **PHStat**. Select **Multiple-Sample Tests→Kruskal-Wallis Rank Test**.

3. Complete the Kruskal-Wallis Rank Test dialog box as shown below. The Sample Data Cell Range is A1 through C11. Click **OK**.

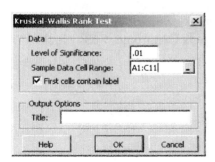

4. Your output should appear similar to the output shown below.

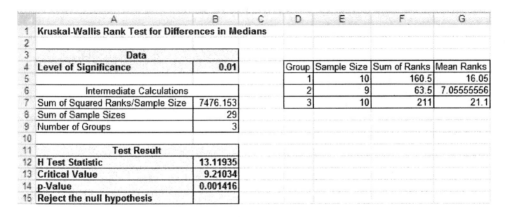

Technology

| ▶ Exercise 3 (pg. 661) | Performing a Wilcoxon Rank Sum Test |

If the PHStat add-in has not been loaded, you will need to load it before continuing. Follow the instructions in Section GS 8.2.

1. Open worksheet "Tech11_a" in the Chapter 11 folder.

2. At the top of the screen, select **Add-Ins** and **PHStat** Select **Two-Sample Tests→Wilcoxon Rank Sum Test**.

3. Complete the Wilcoxon Rank Sum Test dialog box as shown below. You are comparing the Northeast and the South. The Population 1 Sample Cell Range is A1 through A13, and the Population 2 Sample Cell Range is C1 through C13. Click **OK**.

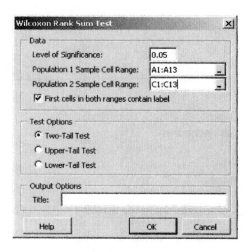

Your output should look similar to the output shown below.

	A	B
1	**Wilcoxon Rank Sum Test**	
2		
3	**Data**	
4	**Level of Significance**	0.05
5		
6	**Population 1 Sample**	
7	Sample Size	12
8	Sum of Ranks	164.5
9	**Population 2 Sample**	
10	Sample Size	12
11	Sum of Ranks	135.5
12		
13	**Intermediate Calculations**	
14	Total Sample Size n	24
15	$T1$ Test Statistic	164.5
16	$T1$ Mean	150
17	Standard Error of $T1$	17.32051
18	Z Test Statistic	0.837158
19		
20	**Two-Tail Test**	
21	**Lower Critical Value**	-1.95996
22	**Upper Critical Value**	1.959964
23	p-Value	0.402504
24	**Do not reject the null hypothesis**	

◀

► **Exercise 4 (pg. 661)**   Performing a Kruskal-Wallis Rank Test

---

If the PHStat add-in has not been loaded, you will need to load it before continuing. Follow the instructions in Section GS 8.2.

---

1.   Open worksheet "Tech11_a" in the Chapter 11 folder.

---

*If you have just completed Exercise 3 on page 661 and have not yet closed the worksheet, click on the **Sheet 1** tab at the bottom of the screen to return to the worksheet containing the data.*

---

2.   At the top of the screen, select **Add-Ins** and **PHStat.** Select **Multiple-Sample Tests→Kruskal-Wallis Rank Test.**

3. Complete the Kruskal-Wallis Rank Test dialog box as shown below. The Sample Data Cell Range is A1 through D13. Click **OK**.

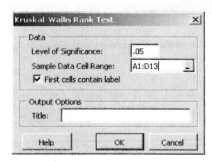

Your output should appear similar to the output shown below. You may have to scroll down a couple rows to see the entire output.

	A	B	C	D	E	F	G
1	Kruskal-Wallis Rank Test for Differences in Medians						
2							
3	**Data**						
4	Level of Significance	0.05		Group	Sample Size	Sum of Ranks	Mean Ranks
5				1	12	311.5	25.9583333
6	Intermediate Calculations			2	12	305	25.4166667
7	Sum of Squared Ranks/Sample Size	29014.46		3	12	251.5	20.9583333
8	Sum of Sample Sizes	48		4	12	308	25.6666667
9	Number of Groups	4					
10							
11	**Test Result**						
12	H Test Statistic	1.032951					
13	Critical Value	7.814728					
14	p-Value	0.79328					
15	Do not reject the null hypothesis						

> ► Exercise 5 (pg. 661)   Performing a One-Way ANOVA

1. Open worksheet "Tech11_a" in the Chapter 11 folder.

*If you have just completed Exercise 3 or Exercise 4 on page 661 and have not yet closed the worksheet, click on the **Sheet1** tab at the bottom of the screen to return to the worksheet containing the data.*

2. At the top of the screen, select **Data** and **Data Analysis**. Select **Anova: Single Factor** and click **OK**.

If Data Analysis does not appear as a choice in the Data ribbon, you will need to load the Microsoft Excel Analysis ToolPak add-in. Follow the procedure in Section GS 8.1 before continuing.

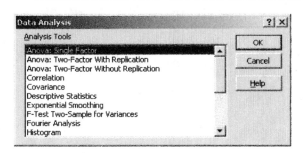

3. Complete the Anova: Single Factor dialog box as shown below. Click **OK**.

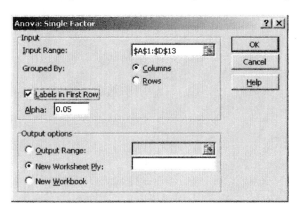

Adjust the column width so that you can read all the labels. Your output should appear similar to the output shown below.

	A	B	C	D	E	F	G
1	Anova: Single Factor						
2							
3	SUMMARY						
4	Groups	Count	Sum	Average	Variance		
5	Northeast	12	377071	31422.58	60802973		
6	Midwest	12	363837	30319.75	38985284		
7	South	12	331435	27619.58	38617798		
8	West	12	364813	30401.08	34585385		
9							
10							
11	ANOVA						
12	Source of Variation	SS	df	MS	F	P-value	F crit
13	Between Groups	95270310	3	31756770	0.734297	0.537163	2.816466
14	Within Groups	1.9E+09	44	43247860			
15							
16	Total	2E+09	47				

PHStat 2.7 with Data Files for use with the Technology Manual to accompany
Elementary Statistics: Picturing the World, 4e
Ron Larson & Betsey Farber
ISBN 0-13-603015-7
CD License Agreement
© 2009 Pearson Education, Inc.
Pearson Prentice Hall
Pearson Education, Inc.
Upper Saddle River, NJ 07458
All rights reserved.
Pearson Prentice Hall™ is a trademark of Pearson Education, Inc.

READ THIS LICENSE CAREFULLY BEFORE OPENING THIS PACKAGE. BY
OPENING THIS PACKAGE, YOU ARE AGREEING TO THE TERMS AND
CONDITIONS OF THIS LICENSE. IF YOU DO NOT AGREE, DO NOT OPEN
THE PACKAGE. PROMPTLY RETURN THE UNOPENED PACKAGE AND
ALL ACCOMPANYING ITEMS TO THE PLACE YOU OBTAINED THEM.
THESE TERMS APPLY TO ALL LICENSED SOFTWARE ON THE DISK
EXCEPT THAT THE TERMS FOR USE OF ANY SHAREWARE OR
FREEWARE ON THE DISKETTES ARE AS SET FORTH IN THE
ELECTRONIC LICENSE LOCATED ON THE DISK:

1. GRANT OF LICENSE and OWNERSHIP: The enclosed CD-ROM ("Software") is
licensed, not sold, to you by Pearson Education, Inc. publishing as Pearson Prentice Hall
("We" or the "Company") in consideration of your adoption of the accompanying
Company textbooks and/or other materials, and your agreement to these terms. You own
only the disk(s) but we and/or our licensors own the Software itself. This license allows
instructors and students enrolled in the course using the Company textbook that
accompanies this Software (the "Course") to use and display the enclosed copy of the
Software on up to one computer of an educational institution, for academic use only, so
long as you comply with the terms of this Agreement. You may make one copy for back
up only. We reserve any rights not granted to you.
2. USE RESTRICTIONS: You may not sell or license copies of the Software or the
Documentation to others. You may not transfer, distribute or make available the Software
or the Documentation, except to instructors and students in your school who are users of
the adopted Company textbook that accompanies this Software in connection with the
course for which the textbook was adopted. You may not reverse engineer, disassemble,
decompile, modify, adapt, translate or create derivative works based on the Software or
the Documentation. You may be held legally responsible for any copying or copyright
infringement that is caused by your failure to abide by the terms of these restrictions.
3. TERMINATION: This license is effective until terminated. This license will
terminate automatically without notice from the Company if you fail to comply with any
provisions or limitations of this license. Upon termination, you shall destroy the
Documentation and all copies of the Software. All provisions of this Agreement as to
limitation and disclaimer of warranties, limitation of liability, remedies or damages, and
our ownership rights shall survive termination.
4. DISCLAIMER OF WARRANTY: THE COMPANY AND ITS LICENSORS
MAKE NO WARRANTIES ABOUT THE SOFTWARE, WHICH IS PROVIDED
"AS-IS." IF THE DISK IS DEFECTIVE IN MATERIALS OR WORKMANSHIP,
YOUR ONLY REMEDY IS TO RETURN IT TO THE COMPANY WITHIN 30
DAYS FOR REPLACEMENT UNLESS THE COMPANY DETERMINES IN
GOOD FAITH THAT THE DISK HAS BEEN MISUSED OR IMPROPERLY
INSTALLED, REPAIRED, ALTERED OR DAMAGED. THE COMPANY
DISCLAIMS ALL WARRANTIES, EXPRESS OR IMPLIED, INCLUDING
WITHOUT LIMITATION, THE IMPLIED WARRANTIES OF
MERCHANTABILITY AND FITNESS FOR A PARTICULAR PURPOSE. THE
COMPANY DOES NOT WARRANT, GUARANTEE OR MAKE ANY
REPRESENTATION REGARDING THE ACCURACY, RELIABILITY,
CURRENTNESS, USE, OR RESULTS OF USE, OF THE SOFTWARE.
5. LIMITATION OF REMEDIES AND DAMAGES: IN NO EVENT, SHALL THE
COMPANY OR ITS EMPLOYEES, AGENTS, LICENSORS OR
CONTRACTORS BE LIABLE FOR ANY INCIDENTAL, INDIRECT, SPECIAL
OR CONSEQUENTIAL DAMAGES ARISING OUT OF OR IN CONNECTION
WITH THIS LICENSE OR THE SOFTWARE, INCLUDING, WITHOUT
LIMITATION, LOSS OF USE, LOSS OF DATA, LOSS OF INCOME OR
PROFIT, OR OTHER LOSSES SUSTAINED AS A RESULT OF INJURY TO
ANY PERSON, OR LOSS OF OR DAMAGE TO PROPERTY, OR CLAIMS OF
THIRD PARTIES, EVEN IF THE COMPANY OR AN AUTHORIZED
REPRESENTATIVE OF THE COMPANY HAS BEEN ADVISED OF THE
POSSIBILITY OF SUCH DAMAGES. SOME JURISDICTIONS DO NOT ALLOW
THE LIMITATION OF DAMAGES IN CERTAIN CIRCUMSTANCES, SO THE
ABOVE LIMITATIONS MAY NOT ALWAYS APPLY.
6. GENERAL: THIS AGREEMENT SHALL BE CONSTRUED IN ACCORDANCE
WITH THE LAWS OF THE UNITED STATES OF AMERICA AND THE STATE OF
NEW YORK, APPLICABLE TO CONTRACTS MADE IN NEW YORK,
EXCLUDING THE STATE'S LAWS AND POLICIES ON CONFLICTS OF LAW,
AND SHALL BENEFIT THE COMPANY, ITS AFFILIATES AND ASSIGNEES. This
Agreement is the complete and exclusive statement of the agreement between you and
the Company and supersedes all proposals, prior agreements, oral or written, and any
other communications between you and the company or any of its representatives relating
to the subject matter. If you are a U.S. Government user, this Software is licensed with
"restricted rights" as set forth in subparagraphs (a)-(d) of the Commercial Computer-
Restricted Rights clause at FAR 52.227-19 or in subparagraphs (c)(1)(ii) of the Rights in
Technical Data and Computer Software clause at DFARS 252.227-7013, and similar
clauses, as applicable. Should you have any questions concerning this agreement or if
you wish to contact the Company for any reason, please contact in writing: Pearson
Education, Inc., One Lake Street, Upper Saddle River, New Jersey 07458 "AS IS"
LICENSE

****
SYSTEM REQUIREMENTS
****
*Microsoft Windows, Windows 98, Windows NT, Windows 2000, Windows ME, or
Windows XP
*In addition to the minimum processor requirements for the operating system your
computer is running, this CD requires a Pentium II, 200 MHz or higher processor
*In addition to the RAM required by the operating system your computer is running, this
CD requires 64 MB RAM for Windows 98, Windows NT 4.0, Windows 2000, Windows
ME, and Windows XP
*Macintosh OS 9.x or 10.x
*In addition to the minimum processor requirements for the operating system your
computer is running, this CD requires a PowerPC G3 233 MHz or better
*In addition to the RAM required by the operating system your computer is running, this
CD requires 64 MB RAM
*Microsoft Excel 97, 2000, 2002, or 2003 (Excel 97 use must apply the SR-2 or a later
free update from Microsoft in order to use PHStat 2.5. Excel 2000 and 2002 must have
the macro security level set to Medium) for Windows; MINITAB Version 14 or 12; JMP
version 5.1; SPSS versions 13.0, 12.0, 11.0; or other statistics software.
*PHStat will not work on the Macintosh
*Microsoft Excel Data Analysis ToolPak and Analysis ToolPak VBA installed (supplied
on the Microsoft Office/Excel program CD)
*CD-ROM or DVD-ROM drive; Mouse and keyboard; Color monitor (256 or more
colors and screen resolution settings set to 800 by 600 pixels or 1024 by 748 pixels)
*Internet browser for Windows (Netscape 4.x, 6.x or 7.x, or Internet Explorer 5.x or 6.x)
and Internet connection suggested but not required
*Internet browser for Macintosh (Internet Explorer 5.x or Safari 1.x) and Internet
connection suggested but not required
*This CD-ROM is intended for stand-alone use only. It is not meant for use on a network.
****
CD-ROM CONTENTS
****
--Data Files : Included within that folder are:
--JMP Data Files (Folder: JMP_Mendenhall_Data)
JMP is required to view and use these files. Information about JMP can be found on the
internet at http://www.jmp.com
--SPSS Data Files (Folder: SPSS_Mendenhall_Data)
SPSS is required to view and use these files. Information about SPSS can be found on the
internet at http://www.spss.com
--MINITAB Data Files (Folder: minitab_Mendenhall_Data)
MINITAB is required to view and use these files. Information about MINITAB can be
found on the internet at http://www.minitab.com/support/index.htm.
--Data Files (Folder: ASCII_Mendenhall_Data)
For use with SPSS or other statistics software.
--Excel Files (Folder: Excel_Mendenhall_Data)
Excel is required to view and use these files. Information about Excel can be found on the
Internet at http://office.microsoft.com/en-us/default.aspx
--TI-83/84 Files (Folder: TI-8x_Mendenhall_Data)
TI-83/84 calculator is required http://education.ti.com/us/product/main.html.
--PHStat 2.5
Prentice Hall's PHStat statistical add-in system enhances Microsoft Excel to better
support learning in an introductory statistics course. http://www.prenhall.com/phstat/
--Readme.txt
****
TECHNICAL SUPPORT
****
If you continue to experience difficulties, call 1 (800) 677-6337, 8 am to 8 pm Monday
through Friday and 5 pm to 12 am Sunday (all times Eastern) or visit Prentice Hall's
Technical Support Web site at http://247.prenhall.com/mediaform.
Our technical staff will need to know certain things about your system in order to help us
solve your problems more quickly and efficiently. If possible, please be at your computer
when you call for support. You should have the following information ready:

- Textbook ISBN
- CD-Rom/Diskette ISBN
- Corresponding product and title
- Computer make and model
- Operating System (Windows or Macintosh) and Version
- RAM available
- Hard disk space available
- Sound card? Yes or No
- Printer make and model
- Network connection
- Detailed description of the problem, including the exact wording of any error messages.

NOTE: Pearson does not support and/or assist with the following:
- 3d-party software (i.e. Microsoft including Microsoft Office Suite, Apple, Borland, etc.)
- Homework assistance
- Textbooks and CD-ROMs purchased used are not supported and are non-replaceable.
For assistance with third-party software, please visit:
JMP Support (for JMP): http://www.jmp.com/support/techsup/index.shtml
MINITAB Support: http://www.minitab.com/support/index.htm
SPSS Support: http://www.spss.com/tech/spssdefault.htm
TI-8x Support: http://education.ti.com/us/support/main.html
Excel Support: http://support.microsoft.com/
PHStat Support: http://www.prenhall.com/phstat/phstat2/phstat2(main).htm
----
Windows and Windows NT are registered trademarks of Microsoft
Corporation in the United States and/or other countries.